CAMBRIDGE
GEOGRAPHY
PROJECT

SOCIETY PIECES

David Lambert

Lecturer in Geography, Institute of Education, University of London

CAMBRIDGE
UNIVERSITY PRESS

Published by the Press Syndicate of the University of Cambridge
The Pitt Building, Trumpington Street, Cambridge CB2 1RP
40 West 20th Street, New York, NY 10011–4211, USA
10 Stamford Road, Oakleigh, Melbourne 3166, Australia

© Cambridge University Press 1993

First published 1993

Printed in Great Britain at the University Press, Cambridge

A catalogue record for this book is available from the British Library

ISBN 0 521 0 40990 X

Designed and produced by The Pen and Ink Book Company
Limited, Huntingdon, Cambridgeshire

Illustrated by Harvey Brazier, Maureen and Gordon Gray,
Andrew Greenwood, Dave Parkins and Pen and Ink.

Cover artwork by Jane Smith

Acknowledgements

The editor would like to thank the following people:
• Alison Bowers for materials and ideas in compiling Unit 4,
  Making Connections
• Claudine Frigère for helpful and needy word processing.

The authors and publishers would like to thank the following:

Fig.2.26 on p.39 and Fig.7.8 on p.125, Crown ©. Reproduced
with the permission of the Controller of Her Majesty's
Stationery Office; Fig.5.10 on p.89 Copyright Plans-Guides Blay
n° 1011; Fig.6.2 on p.102, Central Statistical Office; Case Studies
on pages 115 –116 ActionAid; extract on p.131 from *As I Walked
Out One Midsummer Morning*,© Laurie Lee, 1969, reproduced by
permission of Penguin Books Ltd.

Photographs

Rod Harbinson: p.5; Zefa: p.7, p.20 left and centre, p.23, p.43,
p.58 Fig.3.21, p.63, p.64 right, p.75 C, p.117, p.130 Fig.7.15a,
p.141, p.142 Fig.8.2a, p.145 Fig.8.8b, p.148; Wakefield Art
Gallery: p.9; Network: – Gritsyuk/Matrix p.11; – Sturrock p.51
Fig.3.11; – Pillitz p.63 Fig.4.1c, – Lewis p.63 Fig.4.1d, p.64 left;
– Jordan p.66 Fig.4.5a; – Nona/Rapho p.122; – Matthews p.142
Fig.8.2b, p.146 Fig.8.8d; – Silvester/Rapho p.143 Fig.8.5;
– Goldwater p.144 Fig.8.7, p.145 Fig.8.8a; The J Allan Cash
Photolibrary: p.13, p.16, p.20 right, p.26, p.30 Fig.2.2b, p.56,
p.58 Fig.3.20, p.60 Fig.3.24b, p.66 Fig.4.4b, p.73, p.75D bottom,
p.118, p.125, p.128, p.130 Fig.7.15b, p.133, p.146 Fig.8.8c,
p.150 Fig.8.12; David Lambert: p.14, p.15; Aerofilms Ltd: p.16,
p.25, p.29, p.30 Fig.2.11, p.74; Cheddar Showcaves: p.30
Fig.2.12a; Bournville Village Trust: p.32; London Transport
Museum: p.33; Betty McAskie: p.35, p.41; City Discovery
Centre, Milton Keynes: p.36; Rochdale Local Studies Library:
p.37, p.38, p.39; Milepost 92½: p.51 Fig.3.12, p.63 Figs.4.1a and
b, p.64, p.65, p.75 D right; Sefton Photo Library: p.54, p.66
Fig.4.4a; MetroCentre, Gateshead: p.60 Fig.3.24a; Life File:
– L.J. Hall p.66 Fig.4.4b; – Andrew Ward p.149, p.150 Fig.8.11;
Environmental Picture Library: p.69, p.143 Fig.8.4; Q.A. Photo
Library: p.79 left; Mansell Collection: p.79 right, p.109, p.110
Figs.6.11a and b; Topham Picture Source: p.83, p.101;
Syndication International: p.84, p.110 Fig.6.11c; Barnaby's
Picture Library: p.87; Steve Scoble: p.88, p.89, p.96, p.97, p.98
Fig.5.22; French Railways: p.98 Fig.5.21; Associated
Press/Topham: p.104; David Weight: p.121, p.134, p.137, p.138,
p.139; Leslie Garland: p.126; Yorkshire Dales National Park
Committee: p.129; Mark Edwards/Still Pictures: p.144 Fig.8.6.

Every effort has been made to reach the copyright holders; the
publishers would be pleased to hear from anyone whose rights
they have unknowingly infringed.

CONTENTS

To the pupils

This book is the third in the Cambridge Geography Project series. Book 1, *Jigsaw Pieces*, is concerned mainly with boundaries, frontiers, and the differences between places. Book 2, *Green Pieces*, looks at environmental questions, with a strong emphasis on the physical environments in which people make their lives. This book, *Society Pieces*, mainly deals with people: what people make of their lives. Together these three books cover the whole content of your National Curriculum Geography studies.

More about *Society Pieces*

This book is about *people*. But it is wrong to imagine that people live their lives only as individuals or even only in small groups such as families. People live their lives *in context*. This is firstly their *environment*, and as we found in *Green Pieces*, people can help to make – or damage – their environment. The context is also the *society* in which they live: societies, like environments, can influence people's lives.

For example, look at the front cover of this book: you can identify lots of the different *pieces of society*, like houses, workplaces, transport, communications. These are sometimes called the *infrastructure*. How successful a nation is in supplying these things depends on how its society is organised.

In this book we learn something about the societies in which people live. For example, where should we build new factories? What kind of transport should a city have? How can we encourage people to enjoy the countryside without damaging it? These are some of the questions we ask. A common theme is to ask who gains and who loses from the way things are organised, especially as society *changes*.

The topics include population, settlement, industry, transport, and leisure. The countries we visit include the USA and France, and there is a whole unit on the exciting, economically developing nation of Brazil.

Have fun!

David Lambert

Space and **S**ociety

1·1 *Campaigners protest against the demolition of a historic site in order to build a new motorway link.*

Thinking about the society in which you live

When people live and work together there are always different opinions, which may result in disputes and conflicts. A *successful society* is not one in which there are no disputes; it is one in which the disputes are solved fairly and with broad agreement.

You are an individual. You are also a *member of society*. Your school is an important part of society. Have you ever thought about why we have schools, what they are for, and how they work? If you have, you have been asking questions about society.

One reason why geographers are interested in societies is that societies exist in spaces: in local areas, in your home region, in the nation, a continent, the world. Geographers study the ways in which these spaces are organised by societies.

For societies to work they need an infrastructure of connecting threads: roads, railways, communications.

In groups, think about your home region. You might want to discuss with your teacher what are the boundaries of your home region.

1 Make a list of the different parts of your society.
(*Hint*: Think of all the different jobs people do.)

2 Make a second list of the different parts of the infrastructure that bind the society together and help make it work.

3 On a large piece of paper, arrange your list for (1) in boxes randomly across the whole page. Then draw lines between the boxes to show how particular bits of the infrastructure help to hold the parts of society together. Label each line.

4 Study the letter and the cartoon on the opposite page. On a clean sheet of paper write a heading:
 'Society changes: Shopping'.
a What parts of the infrastructure can you see in the cartoon?
b What parts of society can you see?
c Try to re-draw the scene in the cartoon as a sketch map.
d Draw a line down your sheet of paper. In one column identify members of society who would favour the change described in the cartoon and letter. In the other column identify those people who would be against it.

Members of society who would favour the hypermarket	Members of society who would be against the hypermarket
The reasons for this are:	The reasons for this are:

e What do you think 'market economy' means? Why do you think the person who wrote the letter is against 'market forces'?

5 Using your local knowledge, or referring to a local newspaper, find an example of a similar conflict of space and society. Your example should be a recent or planned change in the way that a part of your society is organised. Find out how the change will affect different members of society.

Setting the scene

Social groups

The human species is a social animal. We like to live in groups. It's good for us to do so because we can share ideas, swap experiences and support each other – all of which makes our lives more comfortable and secure.

Well . . . that's the theory! In fact, living in groups is often quite difficult, because we are all individuals, and we do not always agree all of the time. However, we all have to compete for the means to survive and for a space to live, so it is better to live in a group, working together for these things, than to be solitary.

As a social animal, the human species has developed different societies. These can be very complicated, and consist of many different groups. For example, within the society that is London, well over a hundred different languages are spoken. Many children speak one language at home, and another at school. Some children even speak a third language. This reflects the many different groups to be found in the society.

These different groups are all part of the same society, but their interests are different. Societies develop rules, laws and procedures to help protect the interests of the different groups. For example, in a society like ours that is dominated by cars, it is important that provision is made for minority groups who do not have easy access to cars, for example old or disabled people. These rules and regulations, whether they are local government regulations or national laws, help to shape our society.

The market that forces vegetarianism

MAIDSTONE'S sorry state, highlighted by "Blandsville, UK, 1991" (3 November), is similar to that of most towns left to the tender mercies of developers and the market economy. How is anybody expected to think of these places as "home" any more? My mother has the misfortune of living in the suburbs of Maidstone. She has heart disease, cannot use one arm, and has no car. Her local shop has stopped selling meat. If Sainsbury's in the town centre closes, she will be forced into involuntary vegetarianism. Tell her how choice and quality of life are improved by market forces.

Dr Paul Harris
Eaton, Norwich

1·2 *A letter from the Independent on Sunday, 10 November 1991.*

1·3 *People of many nationalities live and work in London.*

Societies and patterns in small spaces

You probably know your own local area very well. You know that it is not just a jumbled mass of buildings (in a city) or fields (in the country). All the *land uses* form a pattern in the community, though it is sometimes very complicated, and can be difficult to see.

Questions like these will help you to see the patterns in your locality:

- Where are the post boxes? Why are they located in these places?
- Are any types of land use grouped together, such as shops, or offices, or factories?
- In what kind of location do you find car showrooms?

The patterns of land use in spaces do not happen just by accident. There are often good reasons for the patterns we see.

The map on the opposite page shows how Wakefield in Yorkshire grew into an industrial town during the 19th century. Different *residential areas* (where people lived) developed. For example, in 1850 large, 10-bedroomed houses were built in the district of St John's, and back-to-back terraces in York Street. In Nelson Street the small, dark cottages dated back to the 17th century. These residential areas form a pattern.

It is interesting to note the kind of people who lived in these three districts. It is also interesting to learn how the pattern of differences was preserved – and how the fortunate people in the St John's district protected themselves from any change for the worse. The table at the top of this page shows the decisions made by the Building Inspector's Committee. The committee no longer exists, but in the 19th century it was very powerful: it decided what could be built, and where. There were 17 committee members, and eight of them lived in St John's, while others had relatives who lived in St John's.

Records of the Building Inspector's Committee

Type of land use	St John's		Lower York Street		Nelson Street	
	Accepted	Rejected	Accepted	Rejected	Accepted	Rejected
New house	13	0	58	4	2	0
House extension	30	0	59	6	11	2
Office building or shop	0	3	19	0	7	0
Factory	0	6	17	6	6	1

Source: Adapted from K.A. Cowlard (1979) 'The identification of social class areas and their place in nineteenth-century urban development', IBG *Transactions* 4 (2) pp. 239–57.

1·4 *Land use changes in Wakefield, 1856–1901.*

The geographer who studied the evidence in the table says:

'The evidence from Wakefield shows how the land use pattern in the town was preserved for the benefit of the people in St John's. In this way different *social areas* were maintained.'

Examine the table of figures.

1 In what way was the character of St John's preserved between 1856 and 1901?

2 Describe how Lower York Street changed during this period.

3 Nowadays it is the Planning Committee of a local authority that makes land use decisions. Try to find an example in your own area of an interesting land use decision made recently by the local council. Be prepared to tell the rest of the class about it.

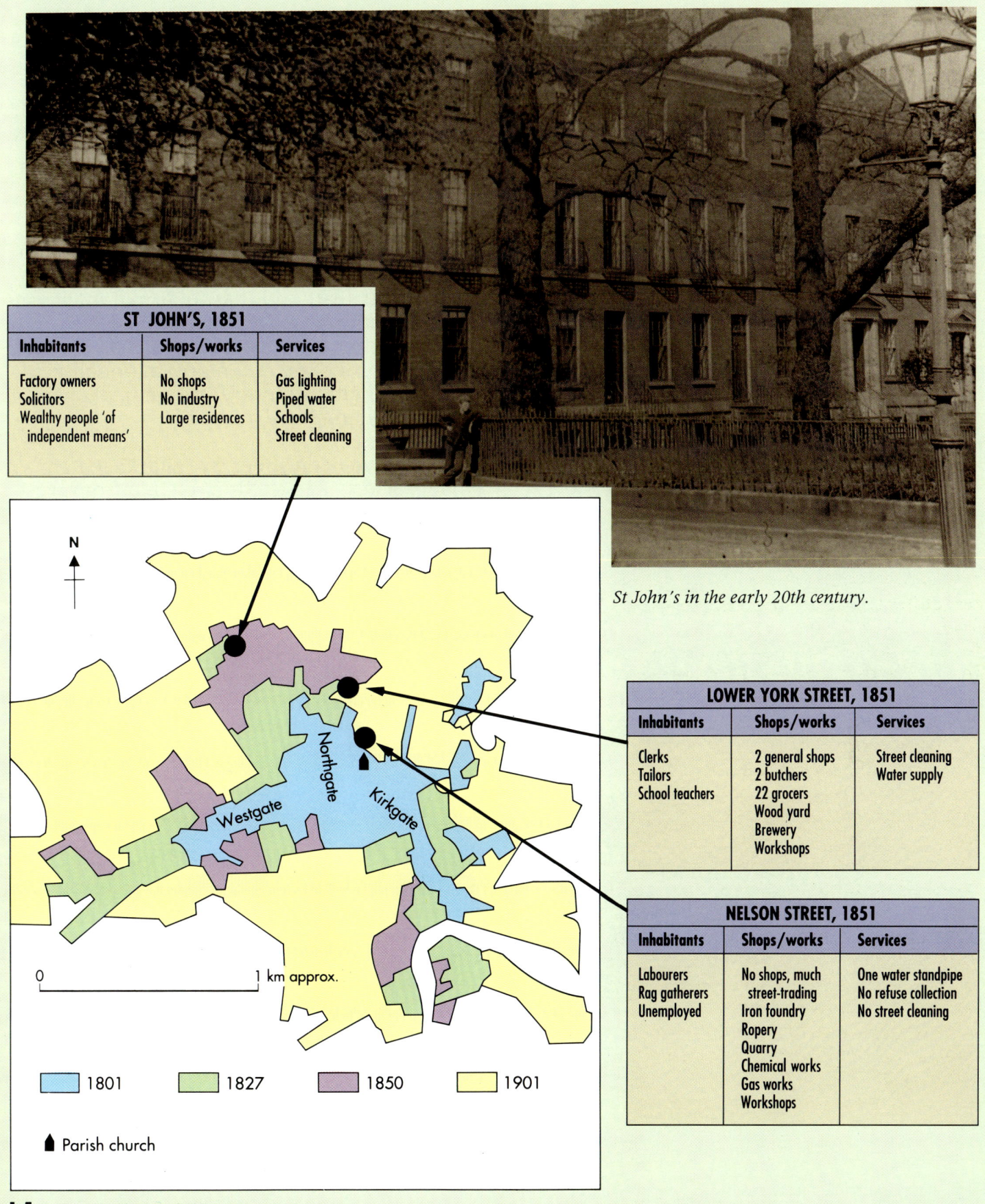

ST JOHN'S, 1851

Inhabitants	Shops/works	Services
Factory owners	No shops	Gas lighting
Solicitors	No industry	Piped water
Wealthy people 'of	Large residences	Schools
independent means'		Street cleaning

St John's in the early 20th century.

LOWER YORK STREET, 1851

Inhabitants	Shops/works	Services
Clerks	2 general shops	Street cleaning
Tailors	2 butchers	Water supply
School teachers	22 grocers	
	Wood yard	
	Brewery	
	Workshops	

NELSON STREET, 1851

Inhabitants	Shops/works	Services
Labourers	No shops, much	One water standpipe
Rag gatherers	street-trading	No refuse collection
Unemployed	Iron foundry	No street cleaning
	Ropery	
	Quarry	
	Chemical works	
	Gas works	
	Workshops	

N

Westgate Northgate Kirkgate

0   1 km approx.

	1801		1827		1850		1901

⏏ Parish church

1·5   *Growth of Wakefield, 1801–1901.*

9

Societies in large spaces

The larger an area, the more difficult it is to hold everything together. Read what a British journalist, living in Washington DC, the capital of the USA, thought when he moved there.

The Big Country

What happens in Washington touches people's lives at few points and, when it does, they are likely to resent it. The most obvious and, therefore, perhaps one of the least remarked features of this country, is its vast size. Within that area is immense variety of climate and landscape, language and culture, tradition and outlook.

The differences are sometimes obscured by such superficial sameness as airport architecture, standard models of cars, and the universal character of Macdonalds. But in reality, the north-eastern seaboard states, say, are as different from California as Germany is from Italy.

Before 1991, the largest country in the world was the Union of Soviet Socialist Republics (USSR). For more than 70 years the communist government tried to create a society that was planned 'from above' by the government. Factories and farms received orders about what to produce, and how much. People of 125 different nationalities in 15 republics were told they must be communists first, and that their nationality was less important. This idea proved impossible for many people to accept. In 1991 the USSR broke up, and many nationalities, led by the Baltic states of Estonia, Latvia and Lithuania, demanded complete independence.

It may have been sheer size that defeated Soviet society. The USSR stretched from west to east over an area equal to half the circumference of the Earth, and spanned 11 time zones – when people are getting up in Moscow, they are going to bed in Vladivostock. But perhaps it was more to do with the mistaken belief that all decisions could be made in Moscow: people in the local areas were made powerless. Certainly, when ordinary people are ignored and have no power to control their own lives, disasters can happen (see Book 2, *Green Pieces*, Unit 6).

1 Find a map showing the countries of the world. Measure the distance across the former USSR from west to east. How many times would the width of Great Britain fit into this distance?

2 What other countries in the world are likely to have difficulties because of their size?

1·6 *From an article by Peter Jenkins, first published on 21 December 1975 in The Guardian.*

1·7 *Pollution from a factory in Russia. Industry can damage your health; because of pollution, life expectancy in Russia declined from 70.4 years in 1964 to 69.3 years in 1990. In some badly polluted cities it is as low as 44 years.*

Society indicators

There are many statistics, or measures, that can be used to indicate the health, wealth or success of a society. *Life expectancy* is one of these. Others include measures of *wealth* (such as income per person), *educational level* (such as the number of people in education after 16 years of age) and *childcare* (such as the infant mortality rate).

Life expectancy is the average number of years a person can expect to live after having survived birth. In the UK it is less for men than for women. This difference could be because more men than women die as a result of accidents. Also, until recently, men smoked more and drank more alcohol than women – these activities can lead to an earlier death.

One of the great challenges for the new Russian state and the other nations that made up the former USSR, is to organise industry and the infrastructure to make a better and healthier life for the whole of society.

The USA faces similar challenges, and others too. For example, the USA is one of the wealthiest nations on Earth and yet 30 million people there live in poverty. Many US citizens ask whether society can be better organised to provide more equal opportunities for all its members.

This book is about *space* and *societies*. We have already looked at some important ideas:

- Societies have patterns, of people and land uses.
- Patterns are always changing.
- Changes nearly always cause conflict.

To help us examine these ideas in more detail there are two key questions to answer in this unit. It is useful to start by thinking of societies in local terms, because we live our everyday lives in small areas. But many of the decisions that shape our local area are not made locally. So the key questions are about space and society on a national scale:

▷ What factors explain the changing patterns of space and society?

▷ How can a society predict future changes?

▷ What factors explain the changing patterns of space and society?

To examine this question, we must first describe a changing pattern. The one we will study here is that of *manufacturing industry* in the USA.

The map below includes the main industrial areas of the north-east USA. These areas include several well-known cities such as Detroit, where Henry Ford first began to manufacture motor cars and which became known as 'Motown'.

At the beginning of the 20th century this region had nearly three-quarters of all the manufacturing jobs in the USA. The region was called *the manufacturing belt* of the USA: the towns and cities offered jobs and opportunities to millions of people and produced goods for the whole nation.

At the end of the 20th century the region is being called *the rust belt*. This name conjures up an image of old closed factories, rusting

NH	New Hampshire
M	Massachusetts
R	Rhode Island
C	Connecticut
V	Vermont
D	Delaware
MD	Maryland
NJ	New Jersey

1·8 *The manufacturing and sun belts in the USA.*

machinery and unemployment. Times have certainly changed. In 1992, General Motors, the biggest manufacturing firm in the world, based in Detroit, announced that it had lost $4,000 million; the company is no longer able to produce the right sorts of cars at the right price for people to buy.

Now, other parts of the USA are more attractive to industry. Most of these are in the south, in an area known as *the sun belt*, which stretches from Florida to California. It is well away from the long, cold, damp winters of the north-east. The proportion of manufacturing jobs in the north-east has declined to less than half.

But it is not all 'doom and gloom' in the north-east. Some of the old industrial areas have taken on a new lease of life. People now work in high-technology industries. This is what Dr John Lambert says about his company, ImmunoGen, in Boston.

1·9 *Smoke billows from a steel mill at Pittsburg, in Pennsylvania, which is in the 'rust belt'.*

ImmunoGen is a biotechnology company that manufactures drugs. After 10 years' research and development (R & D), we have begun to produce a new anti-cancer drug. We started in 1981, renting laboratory space in the university, with one employee. In 1982 we had 5 employees, and by 1987 we had 15. In 1992 we had 165 employees and had begun to manufacture our first drug. Now we have our own labs and factory.

In Sidney Street, where we have our lab, you can see how industry has changed over the years. There are still bits of the old as well as the new. We can see the old Massachusetts Foundry next door – it's still successful because it specialises in high-quality aluminium casings for equipment like computers and hi-fis. Central Fan, further down the road, closed down last year, but The Pipe & Valve Co. and Wheeler Engineering still do OK. But most buildings have been taken over by people like us, or by computer companies like Digital, Wang and Lotus.

There's plenty of competition around here – but that's not surprising. The state of Massachusetts has the greatest concentration of universities and research institutions in the world. It therefore has lots of what high-tech industry needs: brain power. All these highly educated people, and resources like libraries and specialised equipment, attract *venture capitalists*. These are individuals or organisations that provide money for small 'start-up' companies so that new ideas can be tried out. If new ideas can be turned into products, huge profits can be made.

Analysing the changes

Work in small groups. Your task is to find the answers, with evidence, to the following questions.

1 Why did the north-east manufacturing belt develop in the first place?

2 Why did old industries in the north-east decline?

3 Why are sun belt locations attractive to new industries?

4 Why do some cities in the north-east remain attractive to high-tech industries?

You can find evidence by

- looking at the maps and text on pages 12 and 13
- examining the 'Evidence cards' on these two pages
- consulting other books and source material.

1 Old 'heavy' industries

- These were the industries of the first industrial revolution.
- They were established in the late 19th and early 20th centuries. Examples: steelmaking, shipbuilding, chemicals, heavy machinery.
- They used large quantities of raw materials such as coal and iron ore, and fossil fuels as a source of energy.
- In the second half of the 20th century these industries declined in the USA.

2 'King Coal'

- When steelmaking was invented in 1859, about 8 tonnes of coal were needed to produce 1 tonne of steel. (Now less than 1 tonne is needed.)
- Coal is a bulky raw material, which means it is expensive to transport.
- In the 19th and early 20th centuries, bulk carriers had not been invented. It was impractical to transport bulky raw materials any great distance.
- Coalfields, where quantities of coal underground are mined or quarried, became very attractive locations for the old, heavy industries.

3 Electricity

- This is energy made in a power station.
- Much electricity is made from burning a fossil fuel such as coal, gas or oil, although some is now hydro-electric power or nuclear power.
- Electricity was not widely available until the 20th century.
- Although the fuels are bulky, the electricity is not. It can be transported cheaply and easily by cable over large distances.
- The supply of electricity has become part of the essential infrastructure of a country like the USA.

1·10 *An old factory building: Lawrence, Massachusetts.*

4 Consumer industries, light industries and footloose industries

- These are all industries of the 20th century. Examples: cars, household goods like washing machines and dishwashers.
- They use large quantities of *components* which are assembled (put together) in a factory.
- Coal is no longer 'king' for these industries. They can choose from many different locations, as long as they have access to roads and an electricity supply.
- They are no longer tied to raw materials, so are described as 'footloose' industries.

5   Environment and the quality of life

- People who were in charge of light industries or who created new ones, realised that they didn't have to locate their industry in the old industrial towns.
- People like working in an attractive environment.
- Sea, sunshine and mountains give positive images of a good quality of life, and places with a combination of these things are especially attractive, for example California.

1·11   *A new factory: Boston, Massachusetts.*

6   Research and development

- All industries, but especially modern high-tech industries, need research and development (R & D).
- High-tech industries include electronics, computers and biotechnology.
- Men and women who are skilful researchers and who have ideas on how to solve problems or develop new ideas, learn their skills in universities and research establishments.
- Towns and cities with large universities or research establishments are attractive to high-tech industries.
- For these industries, education is now part of the essential infrastructure. That is, education is as important to today's industries as transport and coal were to the old, heavy industries.

The USA – a competitive society?

Still in groups, try this diamond-ranking exercise.

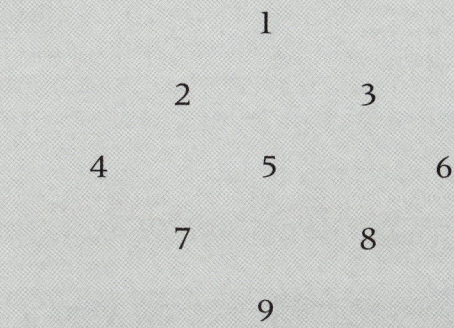

Towns all over the USA compete with each other to attract new industries. Unsuccessful towns slowly die as factories close. Unemployment rises and people move away.

  To attract new industries a town needs certain qualities or *attributes*. Diamond-rank the attributes listed below in what you think is their order of importance.

- A range of other industries already in place.
- Supplies of a fossil fuel nearby.
- Access to a highway (motorway).
- A good infrastructure – schools, shops, services.
- Access to a railway.
- Universities and other research establishments nearby.
- An international airport.
- Flat land to build on.
- An attractive physical environment.

How can a society predict future changes?

In Britain we have a census every 10 years. The most recent census was held in 1991. Every household had to answer a number of questions. A vast amount of information was collected which was processed and then published. You can find census information in the public library. This information is not totally accurate because the government cannot be sure that every single person has been counted. But it is the most accurate record of British society that exists.

It is also useful. It helps us to understand some of the changes that are taking place in society, and to predict changes in the future.

The changing shape of British society

1·12 *Cities in Britain used to be the great magnets attracting people to urban jobs in factories and offices. Now the countryside seems more attractive to many people. Or is it that cities are simply spreading out into the countryside?*

The 1981 census revealed certain trends in British society. Take a good look at these figures.

1·13 *Trends in British society, 1971–81.*

South East	– 1.2
South West	+ 6.0
Wales	+ 2.2
East Midlands	+ 4.8
West Midlands	– 0.6
East Anglia	+11.7
North West	– 2.9
Yorkshire & Humberside	– 0.1
North	– 1.4
Scotland	– 1.8
Great Britain	+ 0.5

Table A: Percentage population change, region by region, 1971–81.

Top four (growing)		Bottom four (declining)	
Milton Keynes	+102%	Liverpool	– 16%
Runcorn	+ 78%	Manchester	– 17%
St Ives	+ 71%	Gateshead	– 21%
Redditch	+ 63%	Salford	– 25%

Table B: Urban population changes, 1971–81.

Figures like these are usually easier to understand if they are displayed in other ways, for example

- by using charts or graphs to display the figures, or
- by using a map.

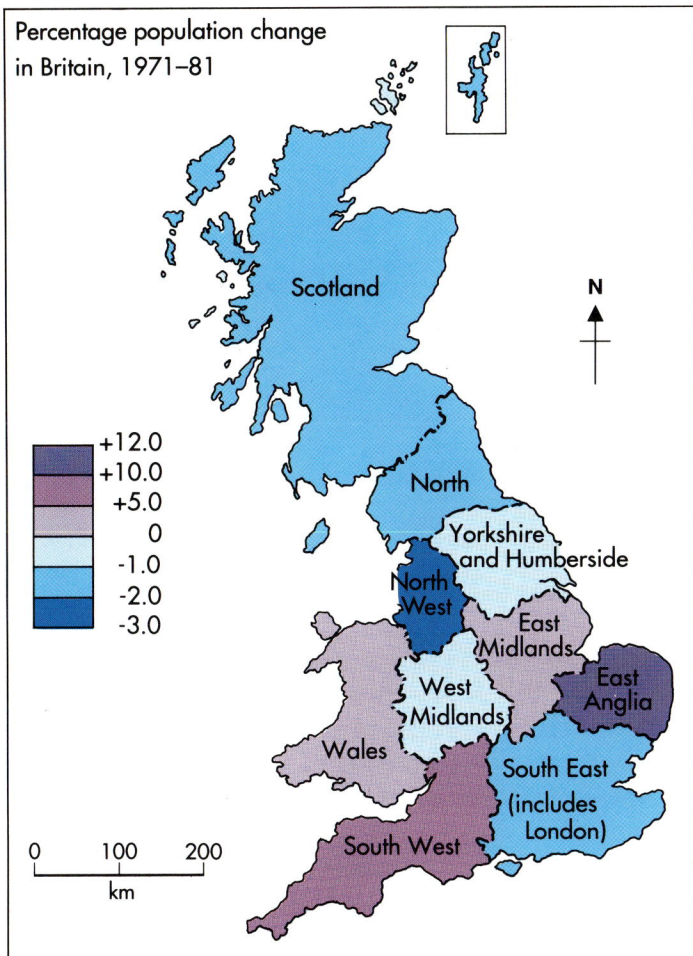

Percentage population change in Britain, 1971–81

+12.0	
+10.0	
+5.0	
0	
-1.0	
-2.0	
-3.0	

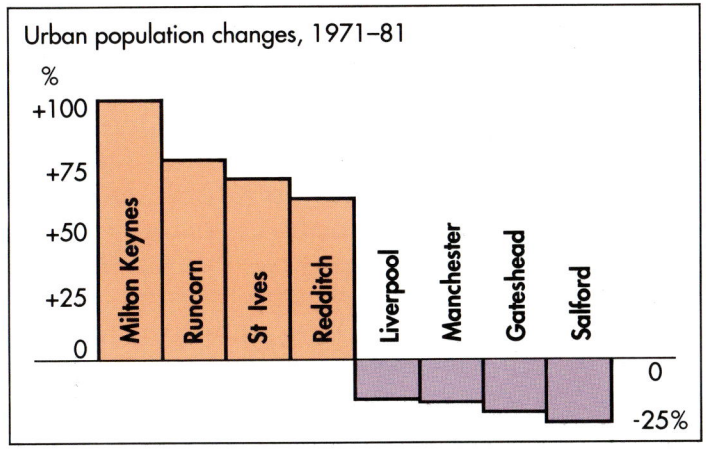

Urban population changes, 1971–81

1·14 *Table A interpreted as a map, and Table B displayed as a graph.*

This is what Professor Linda Murphy said about the figures at the time (in 1981):

These figures show a clear pattern. Regions that are declining are those where there are big cities, for example the urban regions in the South East, and with old manufacturing industries, as in the North and North West. The growing regions are more rural, like East Anglia and the South West. Even more remote rural regions like Wales seem to do well too.

If we just look at urban changes, smaller towns and the new towns have been growing fast. Larger cities, especially in the North, are declining.

Making a forecast

Work through this activity in pairs.

Part One

Study the figures in Tables A and B opposite, and the map and graph on this page. Read again what Professor Murphy said. Imagine you are Professor Murphy and it is now 1990. You have been asked to predict what population changes will be revealed in the 1991 census.

Make *three* different predictions. You cannot just make wild guesses – use your knowledge of existing trends.

Part Two

Now examine the actual results of the 1991 census set out in the tables on the next page. Display the figures in Table C on a map, and those in Table D in a graph.

South East	– 0.1
South West	+ 5.5
Wales	+ 0.3
East Midlands	+ 2.5
West Midlands	– 1.4
East Anglia	+ 7.7
North West	– 4.3
Yorkshire & Humberside	– 1.8
North	– 3.0
Scotland	– 3.5
Great Britain	– 0.4

Table C: Percentage population change, region by region, 1981–91.

Large cities	– 3.6
Small cities	+ 0.6
Industrial areas	– 0.6
Areas with new towns	+ 5.0
Resorts and retirement areas	+ 5.2
Mixed urban/rural areas	+ 3.3
Rural areas	+ 6.1

Table D: Percentage population changes, 1981–91, by type of area.

1a In what ways were your predictions proved correct?
b In what ways were they proved incorrect?

2 It is one thing to describe changes. It is another thing to explain them. Suggest reasons for the changes recorded by the 1991 census.

3 Now make another prediction. What changes to the pattern of UK population would you forecast for the year 2001? The following questions will help you with your prediction:

- Will all the trends you have noticed for 1981–91 continue?
- Will some trends slow down?
- Will some trends speed up?
- Can the government introduce a policy or laws to change a trend, or even reverse it?

The USA grows old

Most countries hold a census from time to time. The US census shows an ageing trend in the population of the USA.

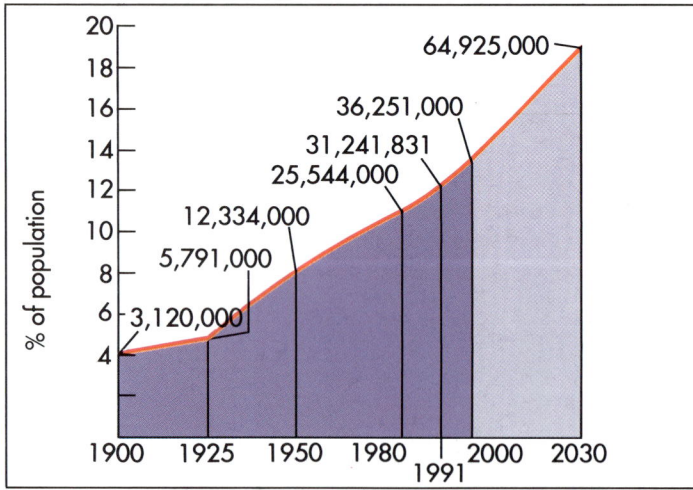

Note: Figures for 2000–2030 are projections (predictions).

1·15 *The population of the USA is growing older. This graph shows the percentage of the total population aged 65 years or more. The average age is also changing. In 1900 the average age of a US citizen was 23 years. In 1985 it was 32, and by the year 2000 it will be 36.*

There are two main reasons for the ageing population: the birth rate has fallen, and the life expectancy has increased. The information cards and illustrations on the opposite page tell you more about these.

A national census provides information about changes. This information helps us to see into the future, to predict future changes. But it does not provide answers to the question, 'What should we do about the changes?' Politicians, our representatives in government, have the power to change things. Adults vote for these representatives according to the ideas they put forward. Ordinary people and groups in society can try to influence the politicians by making them aware of problems, by writing to them, talking to them and making public protests. This is called *lobbying*.

Falling birth rate

This graph shows the changing birth rate in the USA over the second half of the 20th century. The fall in birth rate is the result of millions of women and men deciding to have fewer children. There are theories to explain why they made this decision:

- Children are expensive. Many couples prefer to have a high standard of living rather than a large family. Few (or no) children means that there is more money for holidays, new cars and so on.
- More women and men have decided that a career is more important to them than having a large family.
- There is a greater choice of methods of contraception. Contraception has also become more reliable.

The trend of a lower birth rate is a *social change*. It has consequences for society.

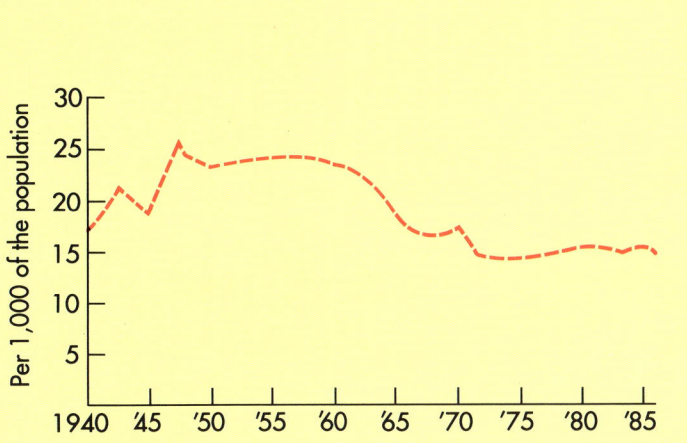

Rising life expectancy

This bar chart shows how life expectancy at birth has changed during the 20th century in the USA. The graph actually hides part of the truth. The fastest-growing age group in the USA is the 'old old' – that is, those who are more than 80 years old. Even at 65, a woman in the USA can expect to live on average another 20 years. These changes are partly a result of medical advances. But they are also a result of social changes. For example, millions of old people have a pension which means that they can eat well, keep warm and live comfortably. This trend also has social consequences.

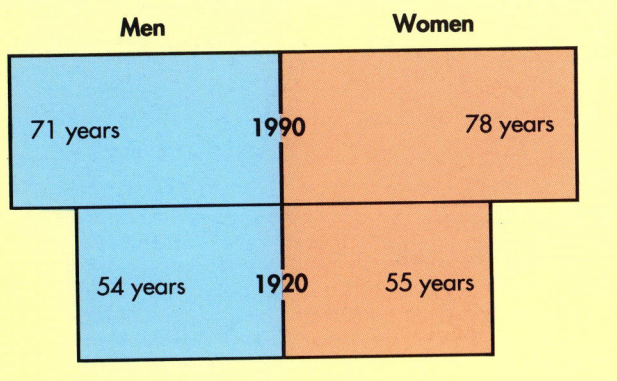

AMERICA GROWS OLD

A report into the consequences of an ageing population
by
Mary Jane Bennett
in Palm Harbor, Florida

Living in Florida is an adventure every day. We have Busch Gardens, Cypress Gardens, Disney World, Epcot, Sea World, and numerous smaller attractions. We have boat rides, even gambling boats that go out to sea before they open their casinos. We have horse racing and dog racing too. We don't participate in these things every day, but it's nice to know they're there to enjoy.

Every day the gardens and shrubs are green and the flowers are beautiful. The weather is a delight, as temperatures are in the 70s and 80s, with only a few cool days in the whole year.

There's no ice and snow – just enough rain to keep things green.
There are hundreds of restaurants within a short distance, all competing with one another, so prices are moderate.

Florida is great, and I wouldn't live anywhere else. But if I'd thought more carefully I wouldn't have moved to a retirement community. So many people seem to be sick or even dying – and it would be good to have some young people around.

1·16 *Living in Florida.*

Work in small groups. You are a lobby group and you live and work in Sun City, a specially built neighbourhood for wealthy, older people in Florida. Choose one of the following jobs:

- bank manager
- gardener
- maintenance worker
- nurse
- shopkeeper
- leisure-centre manager
- doctor
- taxi driver

These are all people who support the people of Sun City – perhaps you can think of others.

Because you live and work in Sun City you think a lot about the needs of old people. You know that there are growing numbers of old people. You also know that many millions of old people in the USA cannot afford to live in places like Sun City.

Your lobby group has come together because you all agree that the US government needs to have a better policy for old people in the future. You plan to launch a campaign, 'The ageing time bomb', to influence both the general public and the politicians. Have a look at the information card 'Policy choices', then work through the following activities.

1 Design *either* a leaflet *or* a poster. It must cover three points:
- What is the problem?
- What could the consequences be?
- What should the government do about it?

2 Write a letter of 100–200 words to accompany the poster/leaflet. Your letter needs to explain your concerns and why you feel strongly about it.

3 Can you think of other policy options that could be added to those listed on the 'Policy choices' card?

4 What percentage of the population in the USA was aged 65 or over in the year you were born?

5 How old will you be in the year 2030?

6 What fraction of the population in the USA will be aged 65 or over in the year 2030?

7 Draw a graph to show the changing percentage of the population aged below 64 years.

8 Read what Mary Jane Bennett has to say about her life in Palm Harbor (opposite page). Would you like to live in a retirement community when you grow old? Explain your answer.

Policy choices

What kind of society do we want? Millions of old people live and die in poverty, partly because they can no longer earn much money. Is this right?

On the other hand, people should plan for their old age by paying pension contributions. But can people afford to save for their retirement, which may last 30 or 40 years?

Some suggestions have been made to solve these problems:
- Build special communities for old people where the things they need can be provided efficiently and cheaply.
- Raise the retirement age to 70 years.
- Introduce a flexible retirement age – let people retire when they want to, or need to.

Summary

The social consequences of an ageing population in the USA can be summarised as follows:

- Greater demand for expensive medical care and hospital services.
- Greater demand for other social services – visiting, meals-on-wheels, etc.
- More expensive pension schemes as people spend a larger part of their lives in retirement, not earning a wage.
- A fall in the ratio between working and non-working people. It was 5 : 1 in 1980; it will be 3 : 1 in 2030. So there will be fewer working people to support the larger retired population.
- The rise of 'grey power': increased political activity by older people.
- Gender imbalance: there are more than twice the number of women than men over the age of 85 years.
- A growth of retirement centres in the sun belt states, especially Florida.

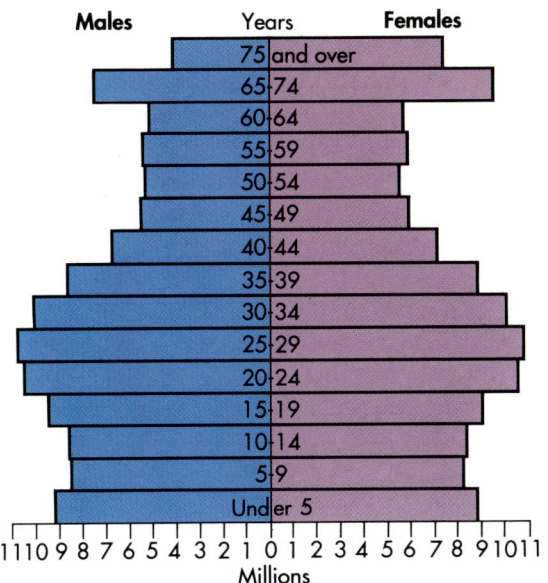

1·17 *Population pyramid for the USA, 1985.*

21

Society

Society is made up of people. The kind of society is determined by how the people make certain choices. For example, how do we ensure healthy, educated people? How much money should be spent on defence, on building roads, providing houses? These are the kinds of question that a society must answer.

Usually societies consist of different groups who have different priorities and make different choices. For example, there are ethnic groups, religious groups, and many other groups within a society.

Space and society

A society makes choices about how to use the space it has. Not everyone in the society agrees, and conflicts arise, resulting in disputes. Successful societies allow conflicts to happen, and encourage people to settle their disputes in a fair way. Part of the work of geographers is to study space and society.

Change

When studying space and society we notice that patterns of land use are usually changing. Perhaps you have noticed where houses have become shops, or shops become offices, where industrial land has become a leisure park, or where farmland has become a golf course. Such changes can be studied to find out who benefits and who loses. In other words, in whose interests do these changes take place?

Research and development (R & D)

Many industries – car manufacturing, food processing, the making of electronic goods – are organised by huge transnational companies on a global scale. Competition is very fierce. To stay ahead, companies must constantly find new products, or better ways of making old products. More and more jobs are to be found in laboratories and research establishments. The people who work there are highly skilled and well-educated people.

Ageing populations

Countries like the USA and Britain, which are sometimes described as 'advanced industrial nations', are facing new circumstances which no society has ever had to deal with before.

People have always grown old, and history shows that all societies, rich and poor, have included individuals who live to 'a ripe old age' – sometimes beyond 100 years old. Recent advances in medicine and health care have ensured that greater numbers of people than ever before now have this opportunity. Fewer youngsters die in childhood from infectious diseases, for example. Old people live longer but need more health care, sometimes for long periods. Just how societies will adjust to these new circumstances is a question that will have to be faced by present and future generations.

Organising **S**ettlements

2·1 *London has been an important settlement for at least 2,000 years. Today, about 7 million people **live** there, and millions more **use** London – as a place to work, for shopping and for entertainment. It is the capital city of Britain, and – like Tokyo in Japan and New York in the USA – a world centre for banks and finance.*

▶ Do you live in a village, a town, a city or in the countryside?

▶ Why do you live where you do?

▶ Do you like living there?

Start-up activity

Part One

1 Draw a circle and write in it the name of the place where you live. Then draw lines from the circle to indicate on one side why you like living there, and on the other why you don't like living there.

The places where people live and work – villages, towns, cities – are called *settlements*. A settlement can be just a cluster of a few families and their houses, or a great city of millions of people. The most important job or *function* of all settlements is to provide homes for people. What are the other functions of *all* settlements?

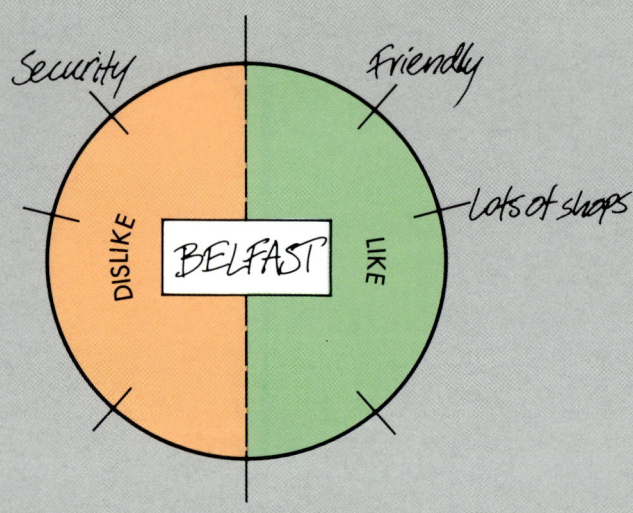

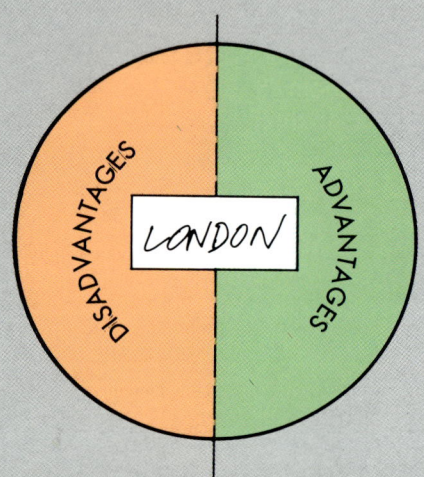

2·2 *Living in Belfast ... and London*

Look at the picture of London on page 23. Does it look to you like an exciting place to live? Many more people live in London than in any other city in Britain . . . but why?

2 Copy the circle with London in the centre, and again draw up two lists. What would be the good things about living in London? What would be the disadvantages of living there?

3 Compare your two diagrams with those of others in the class. How many of you would like to live in London? What are your main reasons?

Part Two

1 Working in groups, draw up a list of the different functions of all settlements. Exchange your list with another group. Which points do you have in common? Did the other group have some different ideas? Make sure each group has a complete list of all the ideas.

There are many jobs done by settlements – but not all settlements do all of them. For example, some settlements have a railway station, so these settlements provide a rail service, both for that settlement, and for other settlements nearby that do not have a station.

2a Decide which settlement you live in (it could be a village, or an area in a town or city).
b Write a list of the jobs done by the settlement in which you live.
c Write a final list of the jobs done in other settlements, but not in yours.

Setting the scene

The jobs done by a settlement are affected by where it is located, and how big it is.

In early Britain, thousands of years ago, most settlements were built for defence. They were usually sited on the tops of hills from where it was easy to see an enemy approaching. To strengthen the defence, ditches or small moats were dug around the edge, and huge mounds of earth and rocks were built up. The remains of these *earthworks* can still be seen today, as at Mam Tor (right).

Defensive settlements were very important in Britain. Later, the Romans built fortresses such as Lindum Colonia (now Lincoln), Verulamium (St Albans), Isca (Exeter) and Vercovicium (Housesteads, illustrated below). Because the Roman soldiers lived in these *garrisons*, they had to buy food and other goods, and civilian settlements grew up nearby, or even inside the fort.

2·3   *Mam Tor, an Iron Age hill fort near Castleton in Derbyshire.*

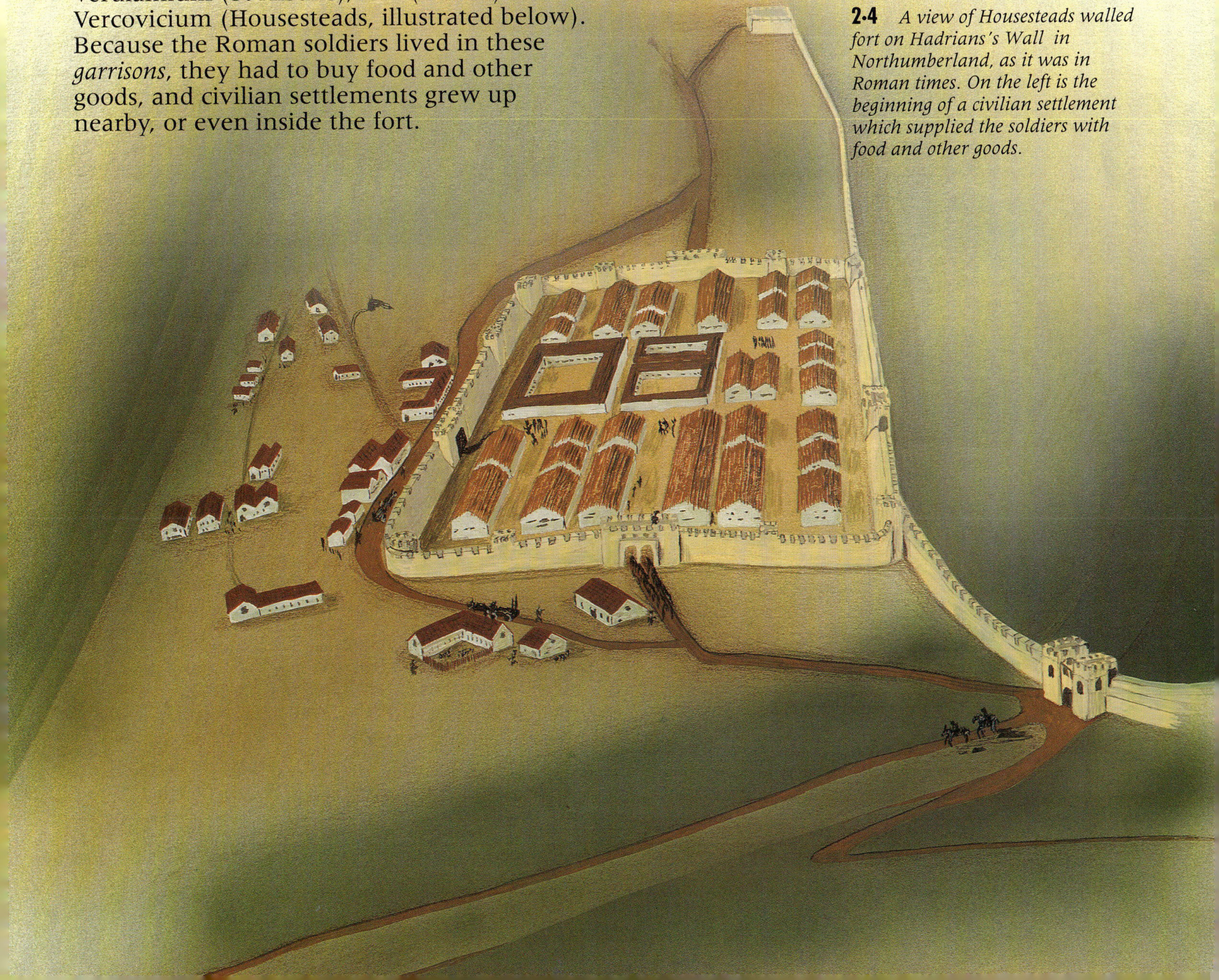

2·4   *A view of Housesteads walled fort on Hadrians's Wall in Northumberland, as it was in Roman times. On the left is the beginning of a civilian settlement which supplied the soldiers with food and other goods.*

2·5 *Conwy Castle.*

New garrison towns were built throughout the Roman, Norman and medieval periods in Britain. They were often sited on the coast, as defence against further invasions from Europe. At Conwy (Conway), on the Wales/England border, Edward I built a castle and a walled town as part of his attempt to conquer the Welsh people. The town walls that we see today (above) are the original castle walls, although some of the buildings within the castle are more recent.

Another reason for the growth of some small settlements was *trade*. Such settlements are known as *market towns*. They developed in agricultural areas where farm produce could be sold or exchanged. They needed good access, so generally they grew up where important roads met, or where there was a ford or a bridge across a river.

Industrial towns began to grow during the industrial revolution in the 19th century. Many of these towns are in the north of England. The first factories were textile mills which were powered by water, so they were sited on the edge of the Pennines where there are fast-flowing streams, in places like Rochdale and Burnley, Bradford and Halifax. Later, when power was supplied by coal-fired steam engines, more industrial towns grew up on and near coalfields. Other settlements grew with the development of the iron and steel and engineering industries – Sheffield and Manchester, for example.

The industrial revolution had a huge impact. In the 19th century, Britain changed from being a nation of country dwellers to a land of industrial towns and cities. In a period of only 100 years, some industrial settlements doubled their size several times over.

	1801	1851	1901
Birmingham	71,000	265,000	760,000
Manchester	75,000	338,000	645,000
Leeds	53,000	172,000	429,000

Source: Gerald Burke, *Towns in the Making*, Edward Arnold, 1977.

2·6 *Rapid settlement growth during the industrial revolution.*

And yet, the space occupied by towns did not increase all that much. It was not until the 20th century that cities really began to spread out in an 'urban sprawl', as society – the people – became more mobile with commuter trains and motor cars. Look at the maps on the opposite page, which show the spread of London.

It is interesting too to look at other changes that were going on in society at this time. While London and other towns and cities were expanding outwards over the countryside, more and more people began to buy their own homes.

Year	Percentage of people owning their homes
1914	10
1939	32
1960	44
1980	55
1990	62

2·7 *Home ownership in Britain, 1914–90.*

▶ Was this trend just a coincidence?

Today, manufacturing is less important. Some of those settlements that were based on industries which have declined, like shipbuilding and textiles, are in decline themselves. Most of these towns are in the north of England, central Scotland, the West Midlands, South Wales and Northern Ireland.

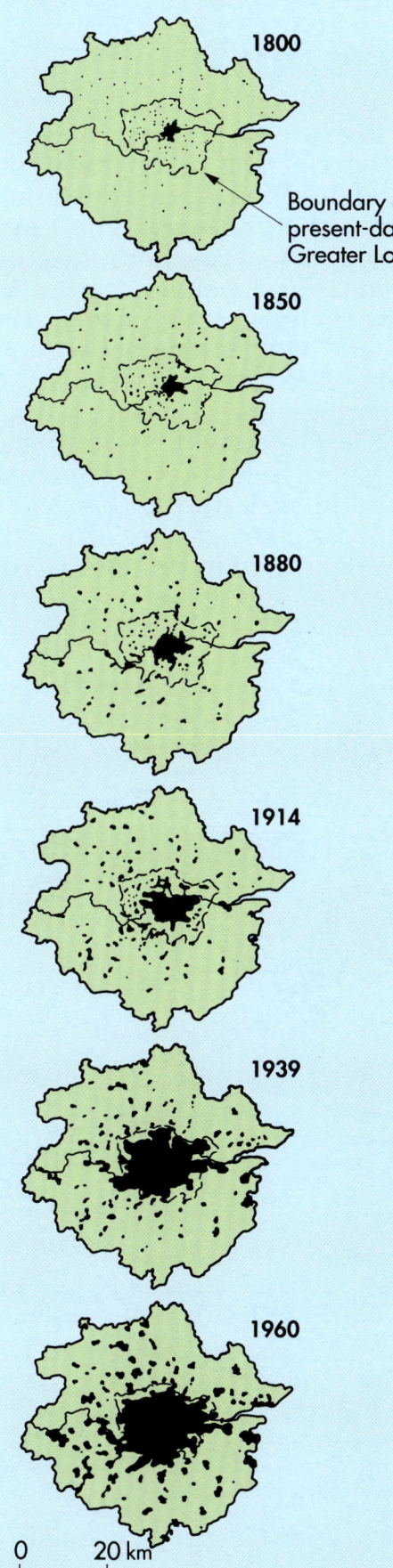

1800

Boundary of present-day Greater London

1850

1880

1914

1939

1960

0   20 km

For example, in the 19th century the Rhondda Valley in South Wales was a busy, thriving coalmining area. Now the last coal mine has closed, and many people are without work. If they are to survive, these people need new jobs.

The settlements that are growing fastest today are those where there are 'high-tech industries', making things like computers and precision instruments. These industries are closely linked with the service industries and with research centres like universities, and are often in an attractive environment. We have already looked at a similar change in the USA (see Unit 1).

Write a report (about 200 words) on the settlement in which you live. It should include answers to the following questions:

- What are the main functions of your settlement? Look back at your notes for Part Two of the 'Start-up activity' (page 24).
- Has your settlement changed over time?
- Is your settlement growing or declining?
- What changes do you predict for your settlement in the future?

In this unit you will look more closely at how settlements in Britain have adapted to all these changes. To guide you there are three key questions:

▷ What are the functions of settlements?

▷ How would you plan a settlement?

▷ How does a settlement cope with change?

2·8 *London spread far and wide, especially during the first half of the 20th century.*

▶ What are the functions of settlements?

The extract from an Ordnance Survey (OS) map (below) shows the settlement of Cheddar and the surrounding area in Somerset. By looking at the map today it is possible to find out about settlements, and learn about the work they do now, and have done in the past.

Most people have heard of Cheddar cheese, but may not realise that the name comes from the village of Cheddar in Somerset. At one time the cheese was made only in this small area, but now it is made all over the world. Clearly, one special job in Cheddar was to make and sell cheese – but Cheddar's history goes back a long way before cheese.

One of the earliest skeletons ever found in Britain was 'Cheddar man'. It was discovered in 1903, and is thought to be of a man who lived during the Iron Age, some 12,000 years ago, when much of northern Britain was still covered in ice at the end of the Ice Age. Today, the skeleton can be seen in the museum at Cheddar, together with even older Stone Age tools and weapons. Gough's Cave, where they were found, is our first clue to the earliest settlement in the area.

The rock here is limestone. Water easily penetrates the rock through cracks. If you look at the map you can see 'dry valleys', where streams used to run, but have now

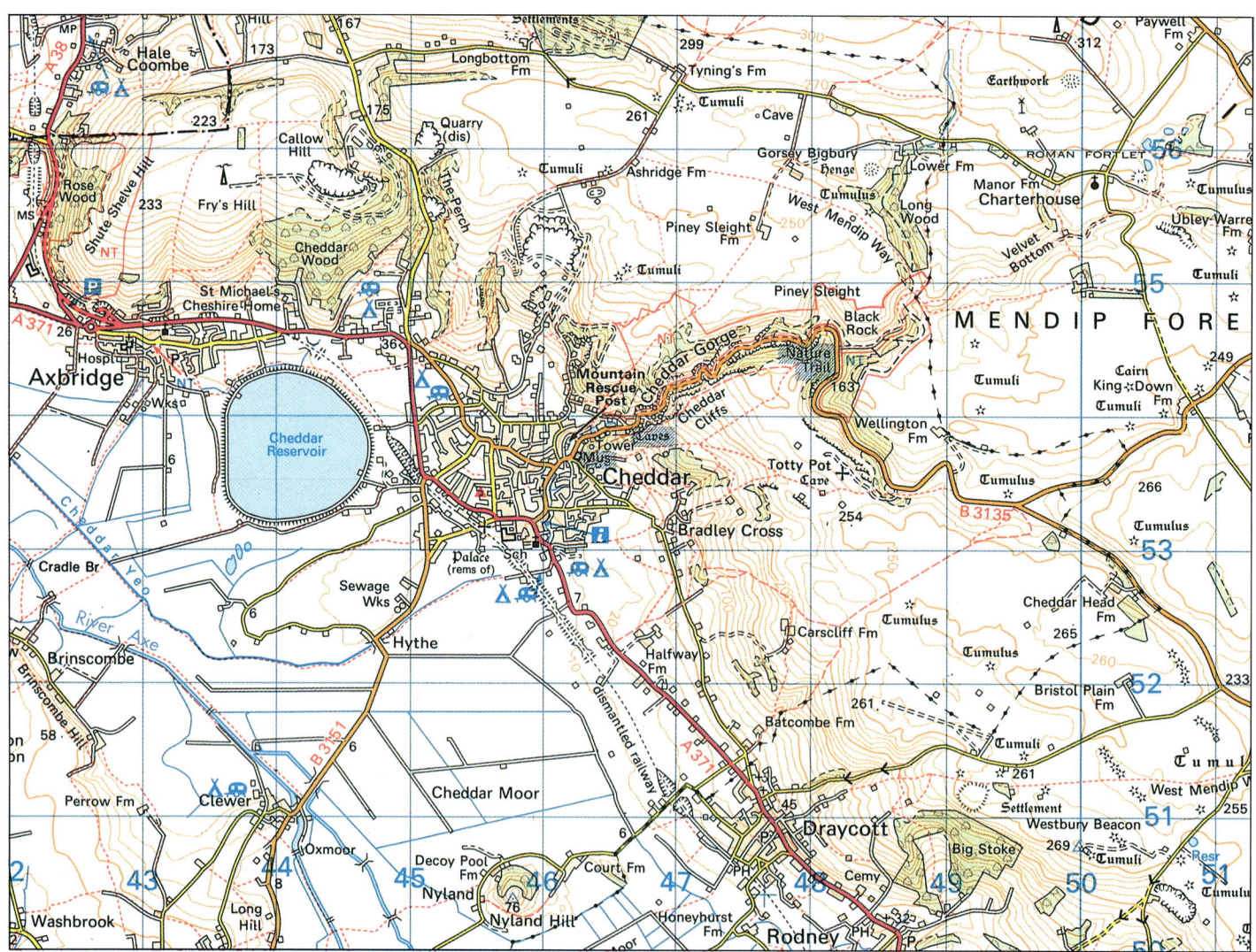

2·9 *A section of the OS 1:50,000 map sheet 182.*

© Crown Copyright

disappeared underground. Below the surface there is a whole network of streams and rivers. Over many centuries the natural acids in the water have dissolved the limestone, making large caves. These provided shelter and water for early people, who used them as natural homes.

In time, people moved out of the caves and built their houses in the open air. These were usually on the nearby Mendip Hills where they could be more easily defended. On the hills they were less likely to be flooded than in the valleys, and it was also easier to travel along the top of the hills, which are fairly flat and open. The valleys would have been filled with forests and swamps. The West Mendip Way (GR 4855) runs along the top of the Mendips. It is now just a footpath, but it was once an important routeway.

Look at the map to find evidence of the remains of early people. *Tumuli* (singular *tumulus*) are earth mounds which are an important clue, but other features are marked too. You can detect them by their medieval style of writing: 𝕿umuli at GR 4656, for example. Are these on the hills or in the valleys?

The Romans lived in these hills too. Iron Age man used the lead that is found here, and this also attracted the Romans. At Charterhouse (GR 4955) there was a Roman mining settlement, with a small fort to protect it.

As people settled and began to farm the land, they gradually cleared the forest and moved off the tops of the hills to work the richer soils in the valleys. But there was still

2·10 *The flat top of the Mendip Hills above Cheddar Gorge.*

the problem of flooding. Look at the valley of the River Axe. It is very wide and flat, and the land was likely to flood from time to time. There was good pasture land here, between the floods, for grazing cattle, which supplied the milk to make the famous cheese – but it was still not a safe place to build a home. Look at the map to see where the villages have grown up. How would you describe their location?

What did the people build their houses with? Limestone is a good building material. Find out what is the OS map symbol for a quarry, then see how many you can find on the map.

What are the functions of settlements today?

People have now controlled the valley area in several ways:

- There are many *drainage channels*, shown on the map as straight blue lines. The land is therefore less likely to flood and provides good pasture for dairy cows and land for crops.
- A large *reservoir* has been built to supply the settlements with water.

Cheese making still provides work for some people, but now they do other jobs too. The 'Wks' on the map (GR 4354), for example, indicates some sort of factory. The quarrying industry is still important too. Much of the quarried stone is crushed into smaller pieces, called *aggregate*, and is used along with sand and gravel in many types of construction ranging from airports and motorways to concrete blocks and footpaths. The quarry at

Callow Hill (GR 4455) and the one north of Cheddar (GR 4655) are both active and contribute to the 9 million tonnes of stone that is *extracted* (taken out) from Somerset each year.

Look at the photograph below and find the area on the map (marked 'Cheddar Gorge' and 'Cheddar Cliffs'). This is a very steep-sided valley and you can see that there is barely space for the road, but it manages to wind a way through the gorge. Cheddar Gorge is a spectacular and unusual feature, which visitors travel far to see. The village of Cheddar encourages visitors, making the most of its natural scenery and early history. You can see symbols on the map to show where there are camping and caravan sites for visitors. Cheddar has found a new job – as a tourist centre.

Use a sheet of tracing paper to make a simplified copy of the OS map of Cheddar on page 28. Mark on it the following details:

- The contour line for 10 metres above sea level. Use a light brown colour to shade in the area above 10 metres.
- Cheddar Reservoir and the main rivers, in blue – but ignore all the fine drainage channel lines.
- The main roads: red for A roads, orange for B roads. Ignore the minor roads which are shown in yellow on the map.
- In black or grey, shade in the main settlement areas of Axbridge, Cheddar and Draycott. Label them clearly on your map.
- Label Cheddar Gorge.

Now decide what else to mark on your map to show what people have used the area for over time. You could mark on the caves, perhaps some ancient tumuli, or some modern tourist centres. You can make up your own symbols, but keep them simple and clear.

- Add a title and a key to explain all the features on your map.

2·11 *Cheddar Gorge from the air.*

a *The gorge.*

2·12 *Cheddar as a tourist centre.*

b *Tourism.*

30

▷ How would you plan a settlement?

Cheddar, and the many towns and villages like it, is an attractive place. Such settlements have grown slowly over the centuries. New buildings were only added a few at a time, and did not destroy the character of the settlement.

On pages 8–9 we read about 19th-century Wakefield. During the industrial revolution many towns like Wakefield grew very fast. To save money, factory owners built the cheapest houses they could for their workers. The houses were packed tightly together, row after row of two-storey terraces with little or no open space between them. Many houses opened straight onto the street and had only small back yards instead of gardens. Parts of many towns became overcrowded and unhealthy.

Charles Dickens was a Victorian writer whose novels about British life shocked many people at the time. They describe the lives of poor people in industrial cities, particularly in east London. You may already know the story of *Oliver Twist*, which highlights the suffering of children who lived on the streets. Read the description below by Dickens, of conditions in London in the 19th century.

Some influential people began to think of ways to protect people from such terrible conditions. They thought about *planning* settlements rather than just letting them grow without any controls.

For example, in 1799, Robert Owen set out to build a 'perfect' industrial village at New Lanark. It was centred around a cotton mill, but Owen believed that the settlement should have other functions too. So as well as building houses for the workers, he also built a school and a co-operative shop.

Slaughter-houses, in the large towns of England, are always (with the exception of one or two enterprising towns) most numerous in the most densely-crowded places, where there is the least circulation of air. They are often underground, in cellars; they are sometimes in close backyards; sometimes (as in Spitalfields) in the very shops where the meat is sold. Occasionally, under good private management, they are ventilated and clean. For the most part, they are unventilated and dirty; and, to the reaking walls, putrid fat and other offensive animal matter clings with a tenacious hold. The busiest slaughter-houses in London are in the neighbourhood of Smithfield, in Newgate Market, in Whitechapel, in Newport Market, in Leadenhall Market, in Clare Market. All these places are surrounded by houses of a poor description, swarming with inhabitants. Some of them are close to the worst burial-grounds in London . . . In half a quarter of a mile's length of Whitechapel, at one time, there shall be six hundred newly-slaughtered oxen hanging up, and seven hundred sheep – but the more the merrier – proof of prosperity. Hard by Snow Hill and Warwick Lane, you shall see the little children, inured to sights of brutality from their birth, trotting along the alleys, mingled with troops of horribly busy pigs, up to their ankles in blood – but it makes the young rascals hardy. Into the imperfect sewers of this overgrown city, you shall have the immense mass of corruption, engendered by these practices, lazily thrown out of sight, to rise, in poisonous gases, into your house at night, when your sleeping children will most readily absorb them, and to find its languid way, at least, into the river that you drink.

2·13 *From an essay by Charles Dickens, 'A monument of folly'.*

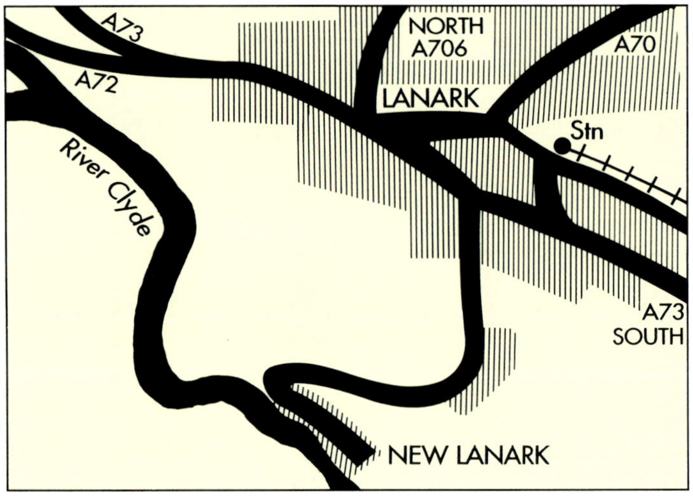

Source: New Lanark Conservation.

2·14 New Lanark, the 'perfect' industrial village.

Nearly a hundred years later, W.H. Lever, who made soap powders, built another new settlement, Port Sunlight, near Liverpool. As well as houses it had a church, hotel, social club and art gallery, bowling greens and swimming pools. Not far away, the Cadbury family, of chocolate fame, built a new factory at Bournville, south of Birmingham. Soon afterwards George Cadbury established a separate housing estate nearby, with unusually big houses and gardens.

These settlements are still there today. But these planned settlements were rare. Most of the town building in the 19th century was unplanned.

Photograph by courtesy of the Bournville Village Trust.

2·15 The Bournville Building Estate in the 1890s. It was unusual at the time for houses to be so spacious, or gardens so large.

Town growth in the 20th century

In the 20th century, transport improved. The railways had continued to expand during the 19th and 20th centuries. Then more and more people were able to buy a car, and for the first time it was possible for people to work in the town then travel home at the end of the day to the *suburbs* on the edge of the town.

Large areas of detached and semi-detached housing sprawled across the countryside. For the individuals who owned these houses it was a major improvement in their quality of life. But then people began to be concerned that some cities would go on and on growing, destroying the countryside and even merging into one another. (Look back at the spread of London on page 23.)

After the Second World War (1939–45) there was a serious shortage of housing in Britain, and it was clear that house building would have to be planned. The idea of setting up a ring of *new towns* around London was put forward – the government would buy land and provide money for building houses, roads and industries in these new towns.

They would attract people away from the congested town centre. Building for the first new towns began in 1949.

Then only a few years later, the idea of the *green belt* was introduced to limit the sprawl of London. Other cities, both large and small, took up the idea. But in many places the sprawl has continued beyond the green belt. The idea of living in small towns or villages is irresistible to people living in the old, overcrowded inner cities.

Green belt

A green belt is land that is protected by law from development. This includes housing and industrial development. The idea of a green belt is to stop urban sprawl, and to prevent settlements from merging into one another, and losing their individual character. The first green belt was set up around London in 1955. Later, other cities decided to have green belts too.

2·16 *Posters like these drummed up business for the Underground railway in London. People were encouraged to buy houses in the new areas which had a different image: not quite countryside, not quite town – the* **suburbs**.

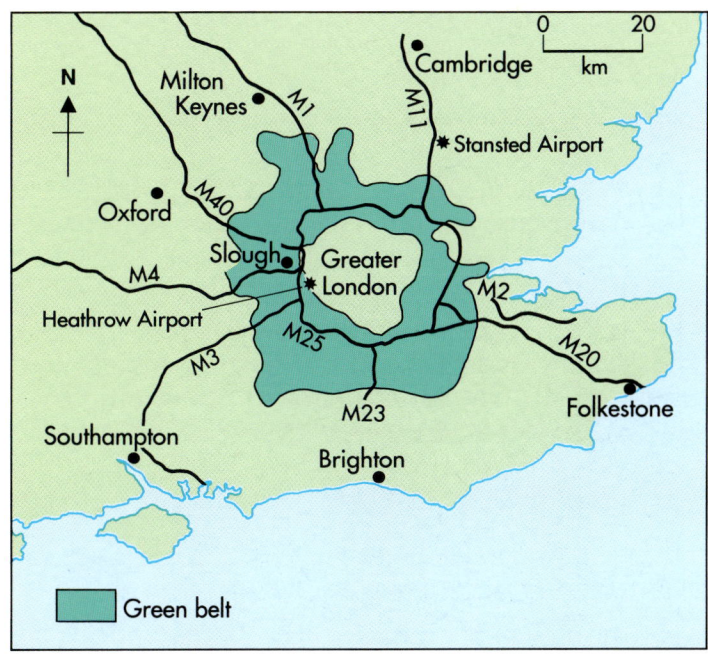

2·17 *London's green belt.*

New towns

The New Towns Act of 1946 allowed the government to buy land and develop it. New towns have been built for two reasons:

1 To provide homes for people who would otherwise live in nearby overcrowded cities.
2 To attract jobs, especially in areas where there is a shortage of work.

The first new towns were set up in the late 1940s and early '50s in a belt around London. They are separated from the city by the green belt but have good rail and road connections with the city centre. There are now 32 government-designated new towns in the UK. There are plans for a number of others, but most of these will not be funded by the government.

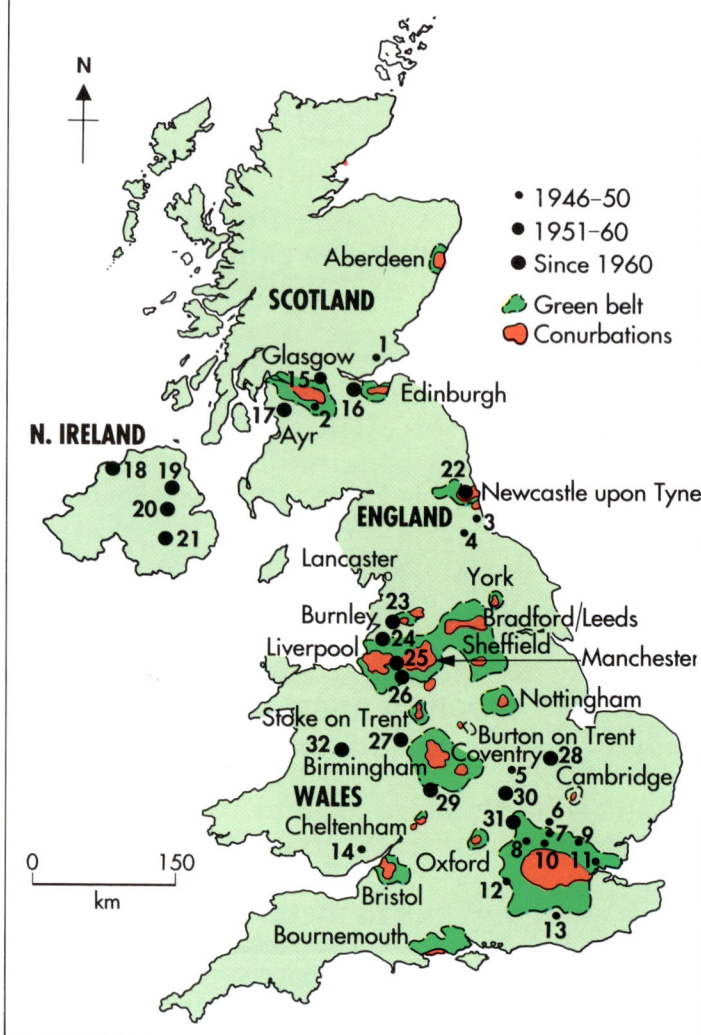

New towns		11	Basildon	20	Antrim
		12	Bracknell	21	Craigavon
1946–50		13	Crawley	22	Washington
1	Glenrothes	14	Cwmbran	23	Central Lancashire
2	East Kilbride			24	Skelmersdale
3	Peterlee	**1951–60**		25	Warrington
4	Newton Aycliffe	15	Cumbernauld	26	Runcorn
5	Corby			27	Telford
6	Stevenage	**Since 1960**		28	Peterborough
7	Welwyn Garden City	16	Livingston	29	Redditch
8	Hemel Hempstead	17	Irvine	30	Northampton
9	Hatfield	18	Londonderry	31	Milton Keynes
10	Harlow	19	Ballymena	32	Newtown

2·18 *New towns, existing and planned, in the UK.*

Work in groups. There is a competition to plan a new town in your area. Each group represents a potential developer. You must draw up a plan, and be prepared to justify your plan to the rest of the class.

First make decisions on the following:

- Do you want a town centre? If so, where?
- What will be the main function of the town centre?
- How will people get around the town? By car? By bus or train? On foot? By bike? What about trams?
- Should you mix up houses, factories and shops? Which should be where? In the centre? On the edge?
- What extra things do you want for your town?

You will also need to show on your plan how your new town is connected by transport links to other towns, cities and regions of the country.

The attempt on the opposite page by one group of pupils is a 'picture map'. You could start your plan in the same way, but your final plan should be in the form of a *land use map*. Discuss how you will do this.

Milton Keynes – special features

- Mainly new housing, from small single-person housing to large family units.
- All the latest architectural designs.
- New ideas in energy-saving homes – some have solar panels.
- All housing is in neighbourhood units, each with its own open space and community facilities such as schools and local shops.
- A large central pedestrians-only shopping area and market, all under cover and surrounded by car parks.
- A range of industrial estates, located away from the houses and close to major roads.

Katie Pope

It's great here, there's so much to do. There are loads of friendly pubs – some are really old because there was a small village here before the city was built. I like the sports facilities best. It's easy to join a leisure centre. In the town centre there's an amazing 'hi-tech pyramid'. Inside is a 10-screen cinema, bars, restaurant and nightclub. There's even a bingo complex – not my sort of thing, though! There are plans for a new ice rink, 36-lane bowling alley and discothèque near the railway station. The only snag here is that buses are expensive. Everything's been built with the car in mind. So you really need a car – or to make sure your friends have!

Scott Williams

This is a terrific place to bring up a family, I sometimes miss old Dagenham, where we came from, but no one else does. Everything here is so clean. There are trees and greenery everywhere. People are friendly – they smile and say good morning. It's so easy to take the kids to school. Tracey, who's 9, goes by herself as there are no roads to cross and she can go with friends. Getting to work is the biggest difference – no traffic jams, and plenty of parking. If we want to visit London, the fastest train takes only 45 minutes.

2·19 *A picture map by a group of pupils.*

What is it like to live in a new town?

Milton Keynes was planned in 1967 to be a major national economic growth centre. It was to be a city rather than a town – the largest settlement ever planned in Britain. It is located in the South East region, which is a general growth area. It was intended to encourage people to move away from London, and so relieve pressure on the capital city. By 1982, the population had reached 100,000, but when Milton Keynes is completed, there should be about 200,000 people living there.

2·20 *Milton Keynes, the new city of the 20th century.*

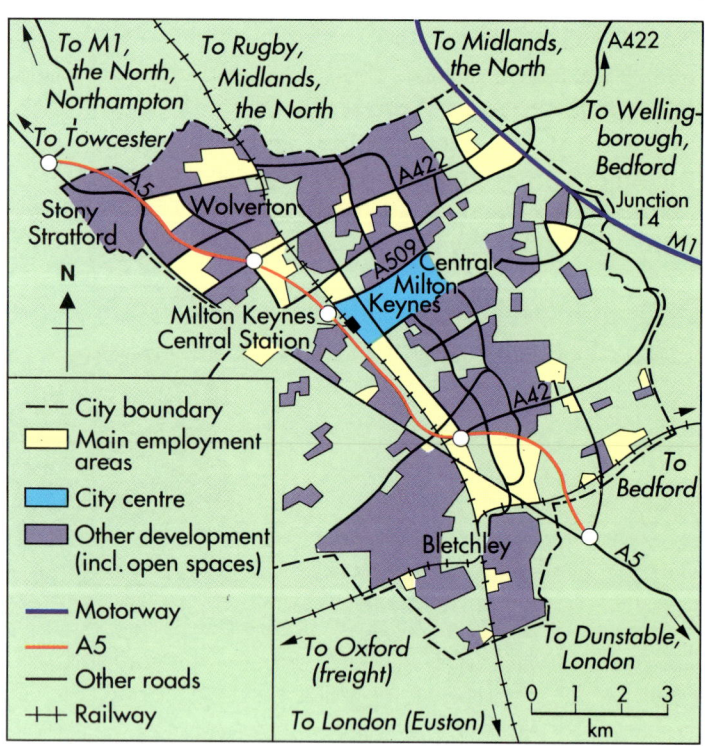

Look at the plan of Milton Keynes and notice:

- the grid-iron pattern of the roads
- the separation of land uses
- the spacious layout
- its location between the M1 motorway and the A5 trunk road.

▶ Is Milton Keynes still the city of the future?

Hold a class or group discussion on this question. You could focus the discussion on these questions:

- In what ways was Milton Keynes planned for the future? (Remember that it was planned in the 1960s.)
- In what ways does the layout of Milton Keynes shape the way people live their daily lives in the 1990s?
- Imagine different futures and how these might affect life in a city like Milton Keynes. For example, imagine a future in which petrol becomes very expensive.

▷ How does a settlement cope with change?

2·21 *Barratt's Yard, Rochdale, in the 1950s.*

Look at this photograph of Rochdale. It was taken in the 1950s. Now look at the picture of another town very near Rochdale, painted by L.S. Lowry, a Lancashire artist, in 1955.

2·22 *'Industrial Landscape', painted in 1955 by the Lancashire artist, L.S. Lowry.*

Reproduced by courtesy of Mrs Carol Ann Danes

▷ What sort of towns were these at that time?

The extract here is from a 1990 tourist leaflet for Rochdale. Can it be for the same place? What is Rochdale *really* like? In this section we learn more about Rochdale, and try to understand how a settlement copes with change.

2·23   *The location of Rochdale.*

Economic change in Rochdale

2·24   *A view of Rochdale in 1780.*

In 1780 Rochdale was a small market town on the edge of the Pennine uplands. Sheep grazed on the hills and the area was known for the spinning of fleeces into thread. Local skills were well established so that when raw cotton was imported from America, Rochdale was well placed to become a cotton spinning town. The streams flowing from the Pennines provided the water power for the early mills, and with the age of steam, coal was available from the Lancashire Plain nearby.

During the 19th century Rochdale thrived and grew, and in this area the main source of prosperity became known as 'King Cotton'. But the new wealth for the few was not without costs. Conditions for the workers in the factories were noisy, cramped and dangerous. Housing for the workers was poor: to save space and costs, terraced

ROCHDALE
Discovering the Pennine Edge

Bordering onto the lower slopes of the hills, Rochdale provides an ideal gateway to the Pennine Edge. Dominated by vistas of open moorland cut by steep, wooded valleys, the area offers a host of secret places waiting to be discovered . . . History and heritage abound in a landscape once frequented by Celts and Vikings. Roman Legionnaires trudged the road over Blackstone Edge and packhorses clattered their way along Reddyshore Scoutgate. The Rochdale Canal and George Stephenson's famous Summit Rail Tunnel gave life to King Cotton, now long departed, leaving only his treasures behind.

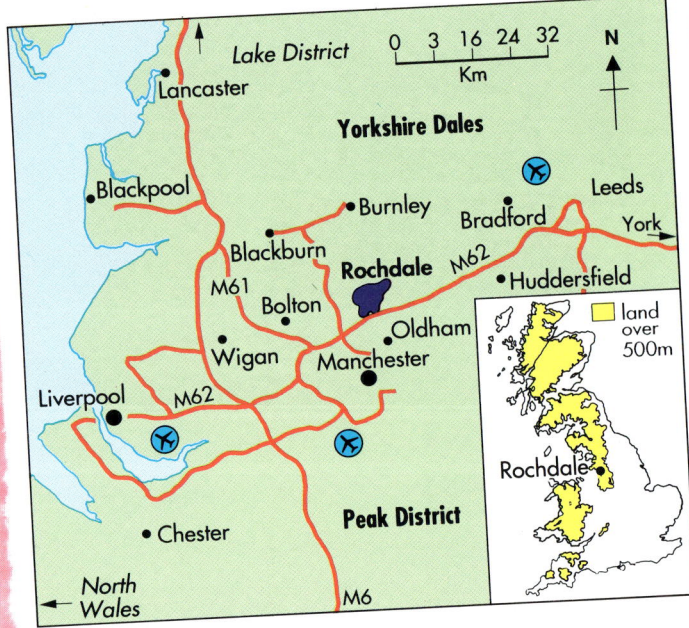

housing was sometimes built 'back to back', so there were not even any back yards.

During the 20th century conditions within the mills slowly improved and better housing was built. Some of the old houses were demolished and replaced by new housing estates, sometimes on the edge of the town. In the 1960s, as in many other towns, high-rise blocks replaced some of the old slums. But Rochdale also led the move to improve the old terraced houses. Deeplish was the first area in the country to be declared a 'General Improvement Area'.

During the 1950s other parts of the world developed their textile industries, such as Japan, Hong Kong and India. The cost of labour in these countries was usually less than in Britain, and the price of fabric and clothing produced there was cheaper than for the same goods made in Rochdale. Not only could overseas factories supply cheaper goods for the world market, but people in Britain – and in Rochdale – were buying them too. Despite campaigns to encourage people to 'Buy British', the sales of goods produced in Rochdale fell.

Rochdale mills began to close. Even Dunlops, once the largest cotton mill in the world, closed and was demolished.

Some of Rochdale's factories will not be mourned. Turners, once the largest maker of asbestos products in the world, was an unhealthy place to work, because asbestos damages people's lungs. But the question of how to stop Rochdale turning into a 'ghost town' became critical. In just five years, between 1976 and 1981, Rochdale lost one-fifth of all its jobs.

a Industries with the largest falls in employment, 1981–87.

Industry	No. employed	% change
Textiles	−1,073	22.5
Mechanical engineering	−959	19.2
Manufacture of non-metallic mineral products	−790	36.4
Food, drink and tobacco manufacturing	−615	38.3
Inland transport (excluding railways)	−538	23.6

b Industries with the largest increases in employment, 1981–87.

Industry	No. employed	% change
Public administration and education	+1,291	14.4
Business services	+790	88.8
Chemical industry	+768	32.2
Wholesale distribution	+713	22.0

Sources: Metro Rochdale Unitary Development Plan (Draft), July 1990, Rochdale Borough Council; HMSO.

2·25 Dunlop's cotton mill, once the largest cotton mill in the world.

2·26 Employment changes in the 1980s.

You have just been appointed as a consultant to Rochdale Borough Council. The Council wants to make Rochdale as attractive as Milton Keynes is, especially to industrialists who create new jobs.

1 In about 100 words, write a short history of Rochdale. This is to be part of a publicity leaflet, which you also need to design, with pictures, a map, and anything else you think is important.

2 Recommend changes that will make the town more attractive to both residents and visitors. The local planners have drawn up a list, and you are asked to rank these items in order of priority. Be prepared to justify your recommendations.

- Clear derelict land, re-using or landscaping it.
- Remove traffic from the town centre.
- Renovate older housing.
- Use the industrial heritage to encourage tourism, e.g. a working textile mill.
- Build a new industrial estate.
- Create a major leisure centre.
- Allocate a site for office development near the town centre.
- Create a new industrial estate near the motorway.

3 Suggest the kinds of industry that the Council should try to attract. Give your reasons. (The tables above will help you with this.)

Social changes in Rochdale

Like many towns in Britain, Rochdale has a very diverse population. It has become a multicultural town. This has happened for *economic reasons*. For example, when Rochdale expanded so fast in the 19th century, there was a shortage of local workers. So factories encouraged people to move from Ireland to work in the Rochdale mills. Then, after the Second World War, many East Europeans came to Britain, particularly from Poland and the Ukraine, which was then a part of the USSR. Some of these people were known as 'displaced persons', because they had been forced from their homes during the War. Others were political refugees fleeing from the political changes in Europe.

In the 1950s and 1960s, when there was fierce competition from textile factories abroad, the Rochdale manufacturers had to try to lower the costs of their goods by keeping the mills open day and night, with people working in shifts. Many local people did not want to work at night, so the manufacturers looked for people who would. Advertisements were placed in Indian and Pakistani newspapers. These were newly independent Commonwealth countries, whose people were pleased to find job opportunities in Britain. The Asian population became the largest *ethnic group*. In 1981 there were nearly 10,000 Asians resident in Rochdale, comprising 10% of the population.

In 1990, Rochdale began planning for the 21st century. In launching its *Unitary Development Plan* it is recognising the diversity of its population, and asks for comments in three languages. Can you identify these languages?

Spinning in the Seventies

Mutual Mills Ltd are at the forefront of Lancashire Spinners for two main reasons: technical experience and marketing experience.

1 Technical Experience

Mutual Mills were one of the first spinners to investigate the spinning of synthetic fibres, starting with Nylon and progressing through to the newer synthetics such as Terylene, Acrilan and Orlon.

2 Marketing Experience

Hand-in-hand with technical ability comes marketing skills, and there is a continued exchange of ideas between every type of textile trade.

Spinning at Mutual Mills is certainly not old – fashioned, but combines traditional craftsmanship with modern know-how and scientific skills. In the field of industrial relations Mutual Mills have long had a reputation for welcoming immigrants of all nationalities. A full list of these nationalities would be impressive, including, as it does, people from China, India, Italy, Malta, Pakistan, Poland, Spain and the Ukraine. Most aspects of induction, organised training and the displaying of notices and posters giving information and facts concerning their jobs and working environment are carried out in at least three languages: English, Italian, and Urdu. Every care is therefore taken to ensure that communications between Management and employees are not hampered by language difficulties.

Mutual Mills Limited

2·27 *This leaflet was printed in 1969. It shows how positively the Mutual Mills viewed their multicultural workforce.*

From 'Immigrants in Industry', December 1969, Local Studies Department, Rochdale Libraries

يہ سب اس بات سے متعلق ہے کہ کونسل کس طرح راچڈیل کے علاقہ میں مستقبل میں زمین کے استعمال اور اس کی ترقی کے لئے پلان بنائے گی اور آپ کس طرح اس کی پلاننگ پالیسیوں پر اثر ڈال سکتے ہیں۔ مزید معلومات کے لئے کس نمبر 514369-0706 پر رابطہ قائم کریں اور یا اگر آپ کو ترجمان کی ضرورت ہو تو آپ انفارمیشن سنٹر، دوسری منزل، میونسپل آفس، اسمتھ اسٹریٹ، راچڈیل پر منگل کے دن دیہہر میں رابطہ قائم کریں۔

Message from Councillor Vernon Earnshaw, chair of the Environment and Employment Committee.

‘ This new development plan will be the first comprehensive plan for the whole borough. It will set the scene for the next decade.

Everyone needs to be aware of its impact on our communities. We are keen to seek as many views and opinions as possible to help decide the contents of the plan.

The major issues are briefly outlined here. We would welcome your comments on these and any other issues so that our plan can take account of all shades of opinion. ’

সমস্ত ব্যারোর জমি জমাকে উন্নত এবং বিশ্য়. ব্যবহার করলে কাউনসিল কি ভাবে কর্মপন্থা নিবার চেষ্টা করিবে সে গুলার উপর আপনি প্রভাব বিস্তার করিতে পারেন ।
অধিকন্তু খবরের জন্য টেলিফন করুন :
-(০৭০৬) ৫১৪৩৬৯ ।
যদি দ'ভাষীর দরকার মনেকরেন তবে ইনফ্রারমেইশন সেন্টা র ২'য় তালা, মিউনিসিপাল বিলডিং, স্মীথ স্ট্রীট, রচডেলে চলিয়া আসুন ।

From 'Metro Rochdale Unitary Development Plan (Draft)', July 1990, Rochdale Borough Council

2·28 *A look to the future – for everyone.*

Prepare items on your home town or region for a noticeboard. Collect local information. Take some photographs and make sketches. Focus on the changes that are taking place.

- Are there changes in employment?
- What sort of jobs are being lost and gained?
- What is the ethnic diversity in your area?
- Where did people come from, and how have they contributed to the character of your home region?

Add ideas of your own to make a lively and interesting display.

2·29 *Rochdale in the 1990s.*
a The new Wheatsheaf shopping centre.
b Children's playground in a residential area.
c An old building housing new businesses.
d A clean, pedestrianised shopping street in the town centre.

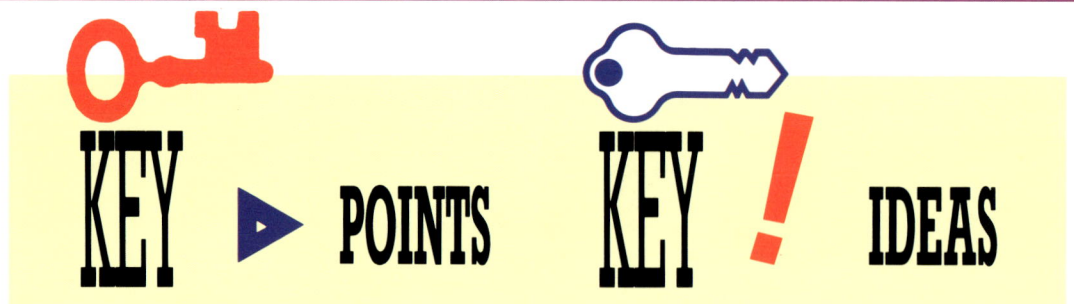

Settlement

A settlement is a place where people live and work. The smallest settlement can be an isolated farmhouse, just one household. A cluster of just a few households is a *hamlet*. A hamlet becomes a *village* when the settlement is big enough to support some basic services: a pub, shop, church and perhaps a primary school. *Towns* are larger still, and the largest settlements are usually referred to as *cities*, though when towns and cities grow and spread into one another they are known as *conurbations*: Merseyside, for example, is a conurbation of Liverpool and the towns around it.

Settlement functions

Settlements have jobs to do. One job, or *function*, is to provide a place to live. The larger the settlement the greater the number of functions it has. Large settlements are like magnets, offering all kinds of shopping, jobs in offices, industries and services, and entertainment, including theatre, cinema, music and sport.

Migration

Towns in Britain grew fast during the 19th century because people migrated or moved in search of jobs in the new industries. Migrants, or their descendants, can tell a rich and fascinating story. Examples of migrants are the Scottish in Corby, Northamptonshire; the long-established black communities in Liverpool; or the Bangladeshi people in east London. The original reason for the migration may have disappeared, but the people stay together. There is enormous potential in the mixing of cultures and experiences for the benefit of future generations.

Change

Like society generally, settlements change. Some grow, others decline. Industrial towns become tourist centres (Bradford). Holiday towns become conference centres (Blackpool). Railway towns become office centres (Swindon). Smaller changes are happening all the time within settlements.

Planning

Long ago people began to realise that many changes could be planned, and the effects of changes would then be more evenly spread. The horrors of unplanned 19th-century industrial towns were enough to convince people of this.

In the 1960s the first ideas were drawn up for the most ambitious planned city ever to be built in Western Europe: Milton Keynes. It is now large enough to have two Members of Parliament of its own. This ensures that a rapidly growing city has adequate representation.

Finding a Place to Work in Britain

3·1 *Trading by computer in the commodities market.*

Thinking about work: making the links

Look at the picture on page 43 and try to decide what kind of work the people are doing. Very few jobs exist 'on their own'. They are usually linked to other types of work in other places. The people in the picture are trading (buying and selling) raw materials or *commodities*, like cocoa or sugar. But there are many other jobs that must be done before they can do their work. They need *information*, answers to questions like:

- How much of the raw material has been produced recently?
- What is its quality?
- What is the production target for the future?

There are many people involved in the answers to these questions – farmers, miners, merchants, lorry drivers, shipping agents, accountants, lawyers, and so on.
  To discover how long the list of jobs linked to the traders in the office can be, see how many jobs you can think of that are related to just one item in the picture: a computer.
  Some of the ideas that one group of pupils had when they were asked this question are shown above (right).

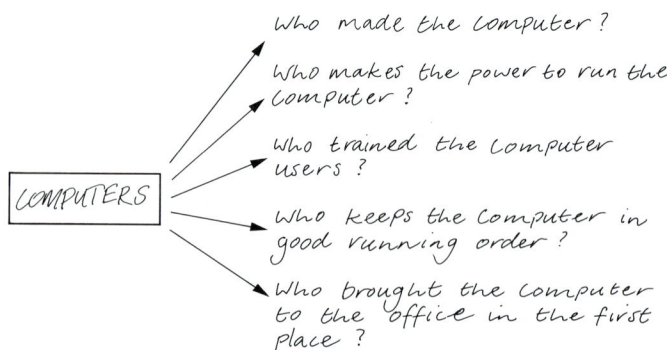

1 Try to think of *other* jobs that are something to do with making, supplying and running a computer.

2 Which of these types of work are *only* linked to computers? Which serve a larger number of people than just the computer users?

3 Now try to increase your list of job-links. Construct a large network 'spider' diagram to display the links you have made.

This activity shows that all the work done in society is related, or linked together. In the UK there is a wide range of different jobs that provide raw materials, finished products, services and expertise to help our daily lives.

Setting the scene

Although there is a wide range of jobs in Britain, they are not equally spread around the country, and are not all open to everyone. Different jobs need different qualities, skills and training. There are more of some types of work in some regions than in others.
  Some work is *manual* and involves mainly physical skills; for example, car workers, mechanics, farmers, hairdressers and drivers are all manual workers. Other types of work are *non-manual* and involve more thinking skills; lawyers, accountants, clerks, teachers and doctors are non-manual workers. We should be clear though that while jobs are described as 'manual' or 'non-manual', these are only very rough descriptions. Many jobs do not fit easily into either of these categories. There are more complicated classifications such as the *standard industrial classification*, which is used in the table on page 45.

Look at these two job advertisements, which appeared in a local newspaper in Coventry.

Jalson's Building Materials plc

Junior Management Trainee

Jalson's require a young, energetic and enthusiastic trainee to work in their administration office.

Qualifications: Maths and English GCSE are essential. Although some exam qualifications are important, it is essential that the person we employ is enthusiastic, able to work in a team, and willing to take on responsibilities and make decisions.

Contact: The Manager, Jalson's Building Materials plc, Tredegar Road, Coventry.

ATKINS SUPERMARKET

Salesperson

Required as soon as possible, a salesperson to work at Atkins Supermarket. Work includes stockroom ordering, stacking shelves, till work and helping customers. Ability in Maths and English preferred.

Contact: The Manageress, Atkins Supermarket, Devon Road, Coventry.

3·2 *Job advertisements in a Coventry evening paper.*

1 What type of job is each of these, manual or non-manual?

2 What skills and abilities does each job require?

3 Look through a national newspaper and find a selection of job advertisements. In what ways are the jobs advertised *nationally* different from those in the local paper?

The unevenness in opportunities for work in Britain may be the result of several factors. For example, one region may have a wide range of jobs, many of them in types of work that are in demand and growing rapidly. Another region may not have many jobs available at all. Or is it that some employers judge people who apply for jobs in their companies in ways that are not always fair? For example, if there were equal opportunities for all people, we would expect employment in *ethnic minority groups* to be the same as for the white population. But if you look at the table below, you can see that this is not the case.

Standard industrial classification (type of industry)	Males		Females	
	White	Ethnic minorities	White	Ethnic minorities
All industries	100	100	100	100
Agriculture, forestry, fishing	3	–	1	–
Energy and water supply	4	–	1	–
Extraction of minerals	5	3	2	–
Metal goods, engineering and vehicles	15	16	5	7
Other manufacturing industries	11	14	10	15
Construction	12	5	2	–
Distribution, hotel and catering	16	27	26	24
Transport and communication	8	12	3	4
Banking, finance, etc.	8	6	10	7
Other services including:	18	17	40	43
Education	4	3	11	6
Medical/health/vet services	2	5	10	17

3·3 *Employment by industry and ethnic origin, 1984–86 (percentage figures).*

Source: Employment Gazette (adapted), March 1988.

1   Look at the figures for white and ethnic minority employment, for males and females, in different types of industry. Are there any areas in which white people are more likely to be employed?

2   In which types of work are ethnic minority women more likely to be employed than white women?

3   In which types of work are ethnic minority men more likely to be employed than white men?

4   Why do you think jobs are not evenly distributed between white and ethnic minority people?

Look at the following passage, which shows what working pay and conditions were like for one Asian mother in Britain.

I work 40 hours a week in the factory and my take-home pay is between £55 and £66 per week. Last year I started a homeworking job which I can do most evenings and weekends. For this I get paid £15–£20 per week depending on the number of overalls I manage to complete. This money adds towards the household budget and occasionally for clothes for the children . . . With the domestic duties and two jobs I have very little time to relax. I don't even have time to fall ill or complain about a backache. I know the work has to be done, as the man would soon come to collect the overalls. My only social life is going to local weddings.

(Mrs P, 42 years old, Asian, with three children)

From C. Oppenheim, 'Poverty: the facts', Child Poverty Action Group, 1990

Now look at these figures:

| | Average weekly earnings | |
	Men	Women
White	£129.00	£77.50
Caribbean	£109.20	£81.20
Asian	£110.70	£73.00

Source: C. Brown, *Black and White Britain*, Third PSI Survey, Gower, 1984.

3·4   *Gross earnings of full-time employees in Britain, 1982.*

These figures clearly show that in Britain in the 1980s, job opportunities and pay were uneven. Women and ethnic minority groups earned much less than white men.

Patterns of industry

Industry

Industry refers to all different kinds of *economic activity.* The term refers to groups of jobs, giving us a broad classification. For example, the phrase *extraction of minerals* includes within it all the different types of mining jobs for a variety of minerals – copper, tin, gold, kaolin, and so on.

Sometimes the term 'industry' is used to refer only to *manufacturing industries* – that is, factories that make products from raw materials.

By dividing up all types of employment into different industries, and collecting figures on the number of people employed in each industry from each region, we can get a good idea of the pattern of work in the country. In Britain these figures are collected each year, and the data can show where there are large gains and losses in employment, and how work patterns change.

Look first at a simple graph of employment for Britain since 1801 (opposite). The jobs have been classified under three headings:

■ **Primary industry**
Mining, agriculture and fishing
These industries produce raw materials.

■ **Secondary industry**
Manufacturing
These industries turn raw materials or components into more useful and valuable finished products.

■ **Tertiary industry**
Services
These are industries that do not make a 'finished product', but provide people with a service, for example a doctor, hairdresser, lawyer, or sales assistant.

It is clear from the graph that the primary industries – mining, agriculture and fishing – now have the smallest workforce of the three categories. This sector has gradually become smaller and smaller. The number of workers in the manufacturing industries is also declining in Britain. The big increase in jobs has been in the service industries. This trend has been continuing over the last 200 years, since the start of the industrial revolution.

The table below shows that, generally, the trends are continuing. But why do you think there was an increase in jobs in the primary sector in the early 1980s? (*Hint*: North Sea.)

	1982	1983	1984	1985	1986	1987	1988
Mining, agric. and fishing	1,756	2,042	1,993	1,953	1,862	1,838	1,843
Manufacturing	5,897	5,525	5,409	5,362	5,227	5,152	5,215
Services	13,406	13,500	13,836	14,108	14,298	14,594	15,168

Source: *Geography* Vol. 75 Pt 4, October 1990.

3·6 *Changes in employment, 1982–88 (millions of jobs).*

We must remember, though, that this is a *national* picture, and that the pattern in each *region* may be different.

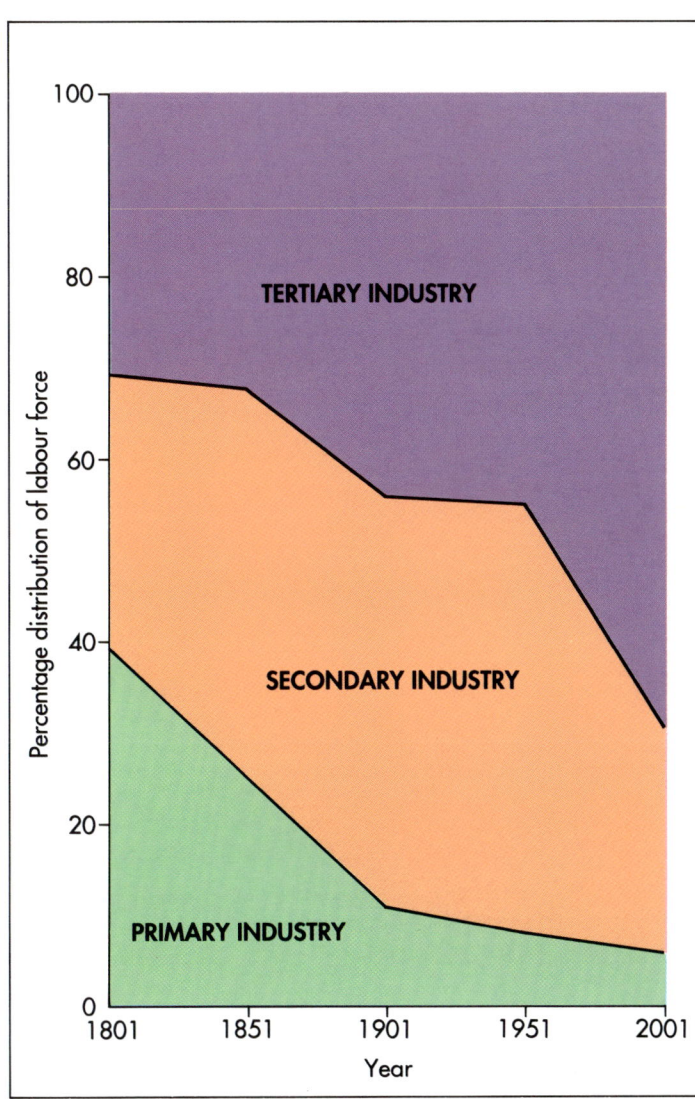

3·5 *Percentage of people employed in the industries in the UK, 1801–2001.*

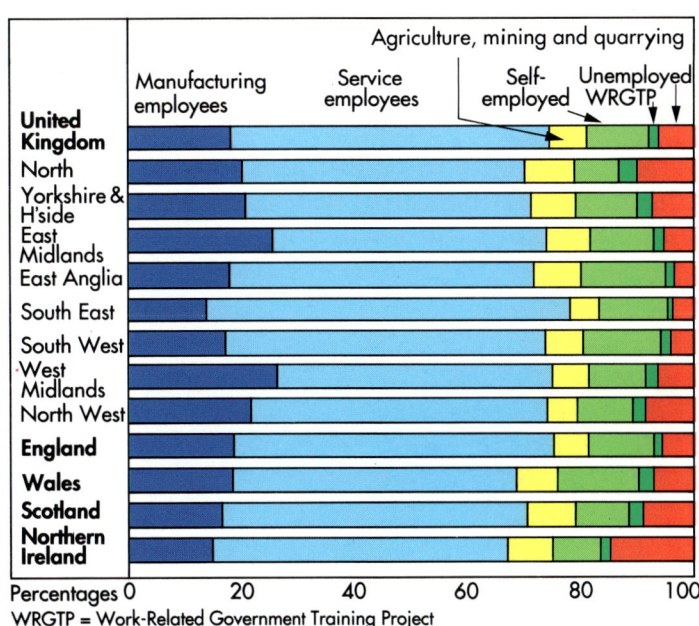

Source: *Regional Trends,* 1990.

3·7 *The UK workforce (national and regional) in 1989.*

The divided bar chart on page 47 adds to the information given in the previous graphs and tables in a number of ways. It not only tells us the proportion of people employed in the whole of the UK (top bar); it also shows these proportions for each of the regions. The right-hand side of the bars shows extra information, such as the numbers of self-employed, those on government work schemes, and the unemployed.

To help us examine all the points raised in this section, we will investigate three key questions in this unit:

▷ What are the jobs that people do?

▷ How does manufacturing industry find a location in Britain?

▷ Where are the service industries in Britain?

Examine all the tables and graphs on pages 45–47.

1 On a base map of the UK, draw bar charts to show the regional variation of manufacturing employment in the UK.

2 Which regions have higher than average employment in manufacturing?

3 Which regions have higher than average unemployment?

4 Can you suggest a reason for any links between your answers to questions (2) and (3)?

5 Which region had the highest percentage of service employees in 1989?

6 By how much did the number employed *nationally* in the service industry change between 1982 and 1988?

▷ What are the jobs that people do?

Work and the home

There are many different jobs to be done in the home. Outside the home, jobs are usually paid, whilst in the home jobs are unpaid.

Domestic work

The table on the next page shows a range of tasks that are done in the home. Make a copy of this table. The list is not a complete one, and you may want to add some items. Think of your own household and try to fill in your table by putting ticks in the correct boxes. If the work is done by a child, put a 'c' in the box as well.

1 What does your table show? Are there some members of the family who are more likely to do certain types of job? Is the work divided up equally between all the family?

2 Can you judge who has the greatest workload in the family? Is this work largely paid or unpaid?

'I'm glad I've got a mum who doesn't work.'

Domestic work – Who Does What?

Type of job	Done by members of the house		Done by paid outside services	
	Male	Female	Male	Female
Housework				
tidying				
cleaning				
hoovering				
washing-up				
cooking				
shopping				
ironing				
House maintenance				
painting				
decorating				
mending broken door or window				
Home improvement				
tiling				
double glazing				
central heating				
plastering				
building an extension				
Car maintenance				
washing				
checking oil				
servicing				
respraying				
Child care				
collecting child from school				
feeding				
bathing				
changing				
seeing doctor, dentist, teacher about child				
child minding				
getting hair cut				

3·8   *Tasks that are done in the home.*

The table you have just completed does not tell the whole story. Also important is the amount of *time* that each person spends on each task, as well as the amount of *movement* that is involved. On the right are timelines for a mother and father in a family in Exeter. The father's timeline has been completed already. Using the information supplied, copy and complete the mother's timeline.

1   Construct work timelines for the members of your family. Compare yours with others in the class.

2   In what ways is 'going to work' less stressful than 'staying at home'?

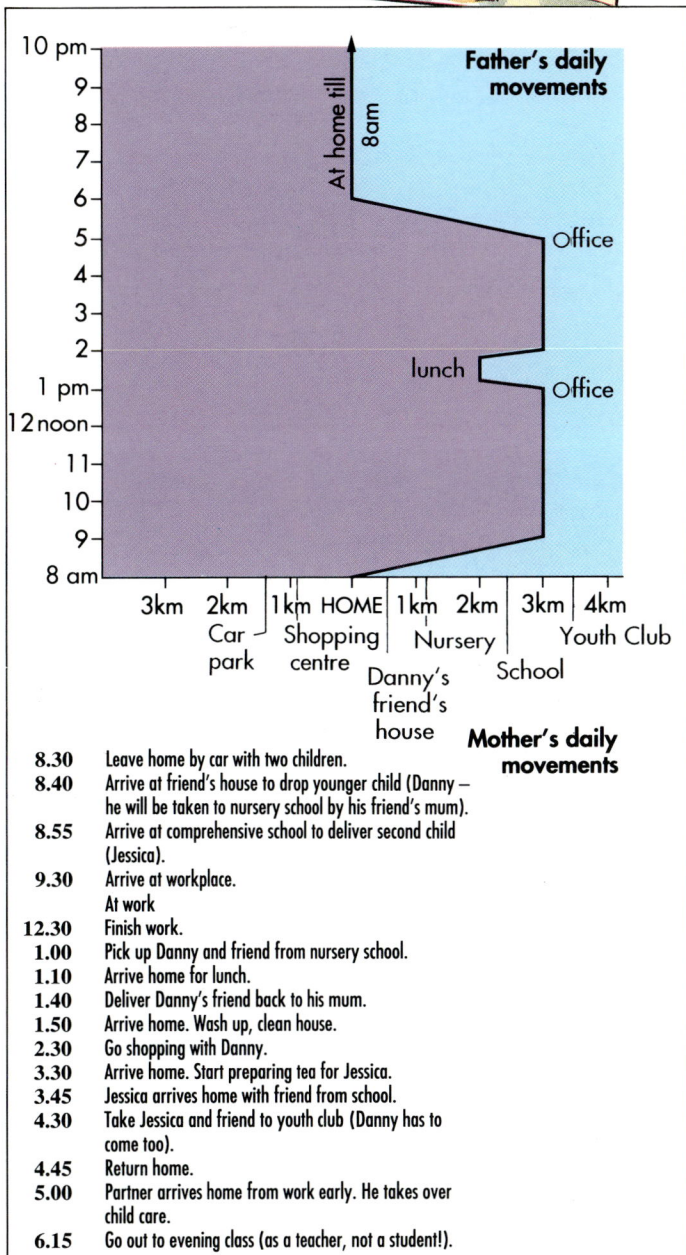

Time	Mother's daily movements
8.30	Leave home by car with two children.
8.40	Arrive at friend's house to drop younger child (Danny – he will be taken to nursery school by his friend's mum).
8.55	Arrive at comprehensive school to deliver second child (Jessica).
9.30	Arrive at workplace.
	At work
12.30	Finish work.
1.00	Pick up Danny and friend from nursery school.
1.10	Arrive home for lunch.
1.40	Deliver Danny's friend back to his mum.
1.50	Arrive home. Wash up, clean house.
2.30	Go shopping with Danny.
3.30	Arrive home. Start preparing tea for Jessica.
3.45	Jessica arrives home with friend from school.
4.30	Take Jessica and friend to youth club (Danny has to come too).
4.45	Return home.
5.00	Partner arrives home from work early. He takes over child care.
6.15	Go out to evening class (as a teacher, not a student!).

3·9   *Daily movements of a mother and father in a family.*

Work outside the home

In 'Setting the scene' on pages 44–46 we saw different ways to classify types of work. Here, Professor Doreen MacDougal talks about a further classification:

A useful classification is the one that divides work into *primary*, *secondary* and *tertiary* jobs. But no classification is perfect. For example, the last 10–20 years have seen the rapid growth of a very special type of industry: Research and Development (R & D). It is usually lumped in with tertiary jobs. But these are not just 'services'. They're special. We now call them *quaternary jobs*.

1 Put each of the following job descriptions under the correct heading: *primary, secondary, tertiary, quaternary.*

footballer	salesperson
banker	actress
coal miner	software engineer
hairdresser	electrician
teacher	forensic scientist
car maker	steelworker
police constable	research engineer
politician	mechanic
biochemist	beautician
farmer	secretary
telephonist	

2 Now look at the map and table on the right. (The quaternary sector is still small, so has been included with the tertiary figures.) Copy or trace the four proportional circles that are left blank on the map, then use the figures in the table to label each circle with the correct figures.

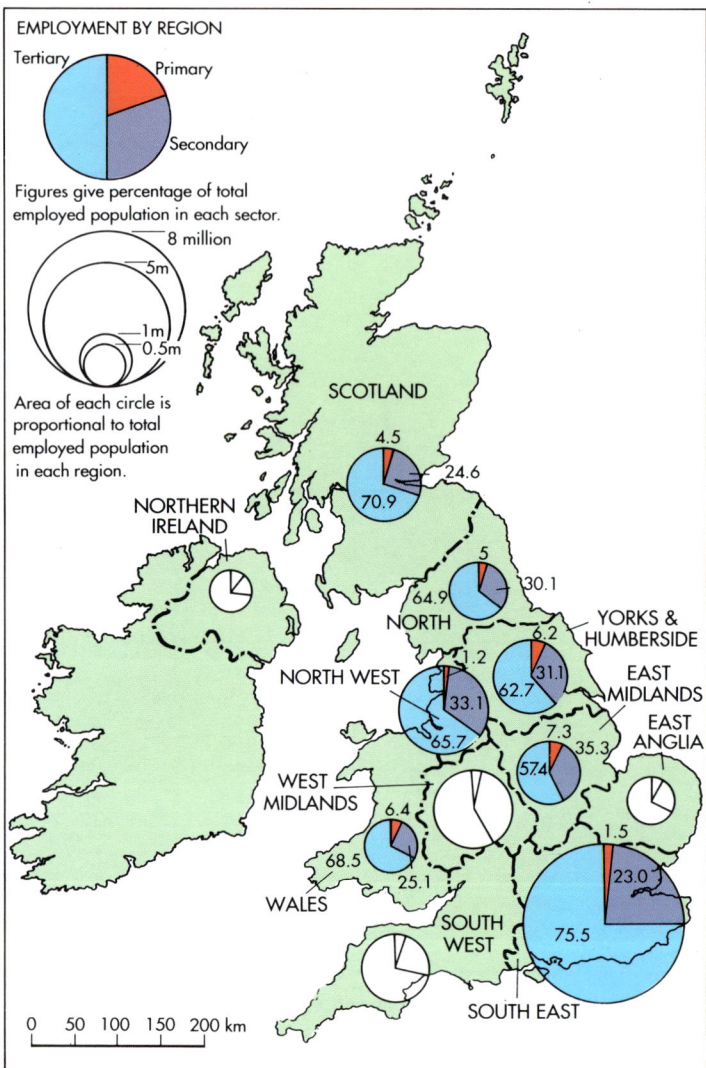

EMPLOYMENT BY REGION

Figures give percentage of total employed population in each sector.

Area of each circle is proportional to total employed population in each region.

	Primary	Secondary	Tertiary
North	5.0	30.1	64.9
Yorkshire & Humberside	6.2	31.1	62.7
East Midlands	7.3	35.3	57.4
East Anglia	7.3	27.6	65.1
South East	1.5	23.0	75.5
South West	4.5	24.5	71.0
West Midlands	3.2	38.4	58.4
North West	1.2	33.1	65.7
Wales	6.4	25.1	68.5
Scotland	4.5	24.6	70.9
Northern Ireland	5.7	22.3	72.0

Source: Walford *et al, A Geography of Contemporary Britain,* Longman Group UK Ltd, 1990.

3·10 *Percentage of people employed in each type of industry in the regions of Britain, 1989.*

Changing jobs

The map below shows rates of unemployment in Britain. Most of the people who are described as 'unemployed' are changing jobs, sometimes because their old job has disappeared. People employed in old-style manufacturing industries like shipbuilding have found it very difficult to find new jobs suitable to their skills. In these circumstances unemployment can become a long-term prospect.

People cannot afford to stay unemployed for too long if they want a reasonable standard of living. They may be forced to work far away from the region where they live. Often this means travelling away from home – commuting – for a week at a time. Look at the passage below.

The last ship launched from Smiths Dock, near Middlesbrough. Unfortunately it has been difficult for shipworkers to find work in new industries or service industries.

The scene at Lime Street Station, Liverpool, around midnight on a bone-numbing Sunday had the time-warped feel of an old newsreel: a long line of shabbily-dressed men shuffling and stamping against the cold, stirring jerky black-and-white memories of civil upheaval, a stream of refugees.

But then these people are refugees, refugees from joblessness, making their weekly way to work in a place where they cannot afford to live, London.

These people work mostly in London's ever-lively building industry; a few work below stairs in the capital's hotels and restaurants. During the week they squat, live in huts on site or sleep in bed and breakfast joints in the East End and the inner suburbs. On Friday nights they take the train back to Liverpool for two days with their families before returning on the cheap fare midnight mail train.

From 'Scousers: doing the South's dirty work' by C. Nevin, in the Daily Telegraph, 19 January 1987

SCOTLAND

NORTHERN IRELAND

NORTH • Middlesbrough

NORTH WEST
Liverpool •

YORKSHIRE & HUMBERSIDE

EAST MIDLANDS

WEST MIDLANDS

EAST ANGLIA

WALES

London •

SOUTH EAST

SOUTH WEST

8% and above
6–7%
Below 6%

3·11 *Rates of unemployment in 1989.*

3·12 *The way of life for a weekly commuter. Commuters arriving at Waterloo station, London.*

▷ How does manufacturing industry find a location in Britain?

Locating manufacturing industry in the 'right' place is very important. When a decision is made to build a new factory at a particular place, there are a number of points to consider. These concern the things that are needed to make the factory work – sometimes called *factors of production*.

Look at the list below. These are the main things needed to run a factory. They are not always needed in the same quantity – for example, some factories only need a few workers while others need thousands.

Power – often electricity.
Labour – people to work in the factory.
Land – somewhere to site the factory.
Transport – a means by which to get raw materials into, and finished products out of, the factory.
Money – to buy raw materials, pay the staff, lease the building, buy machinery, advertise the products, and so on.
Raw materials – the materials, or components, from which the finished product is eventually made.

All of these items are *inputs*, because they all go into making the factory work.

1 What do you think the outputs of a factory might be? Copy the diagram below, and try to label all five of the 'output' arrows.

2 Could a factory work successfully if one or more of the inputs were not available? Why?

3 Transport costs are extremely important to some industries. Carrying the raw material from its source to the factory, and then taking the finished product to where it will be sold (the market) can be expensive.
a Where will a factory try to locate if the cost of transporting the raw material is much greater than transporting the finished product?
b Where will a factory try to locate if the cost of transporting the finished product is much greater than transporting the raw materials or components?

4 Why can a factory not always be built exactly where the owners would like to build it?

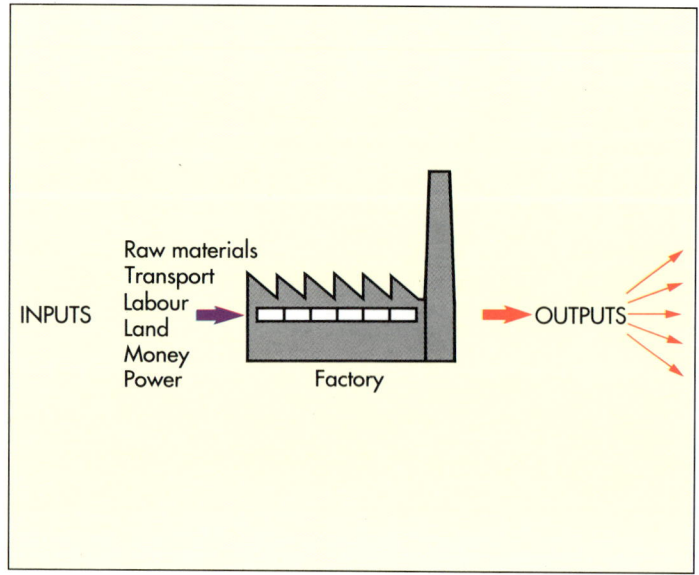

3·13 *Factory inputs and outputs.*

Decision making

Work in small groups of three or four. Look at the map opposite, which shows an area where some industrialists (people who organise and run an industry) are thinking of locating a factory. They have put together a table of important information to help them decide where to build the factory. The points that would save most money are given the highest scores.

The map gives you a choice of five possible sites. Using the scores in the table below the map, work out which site is the best for the industry to locate in. Fill in a copy of the blank table to help you.

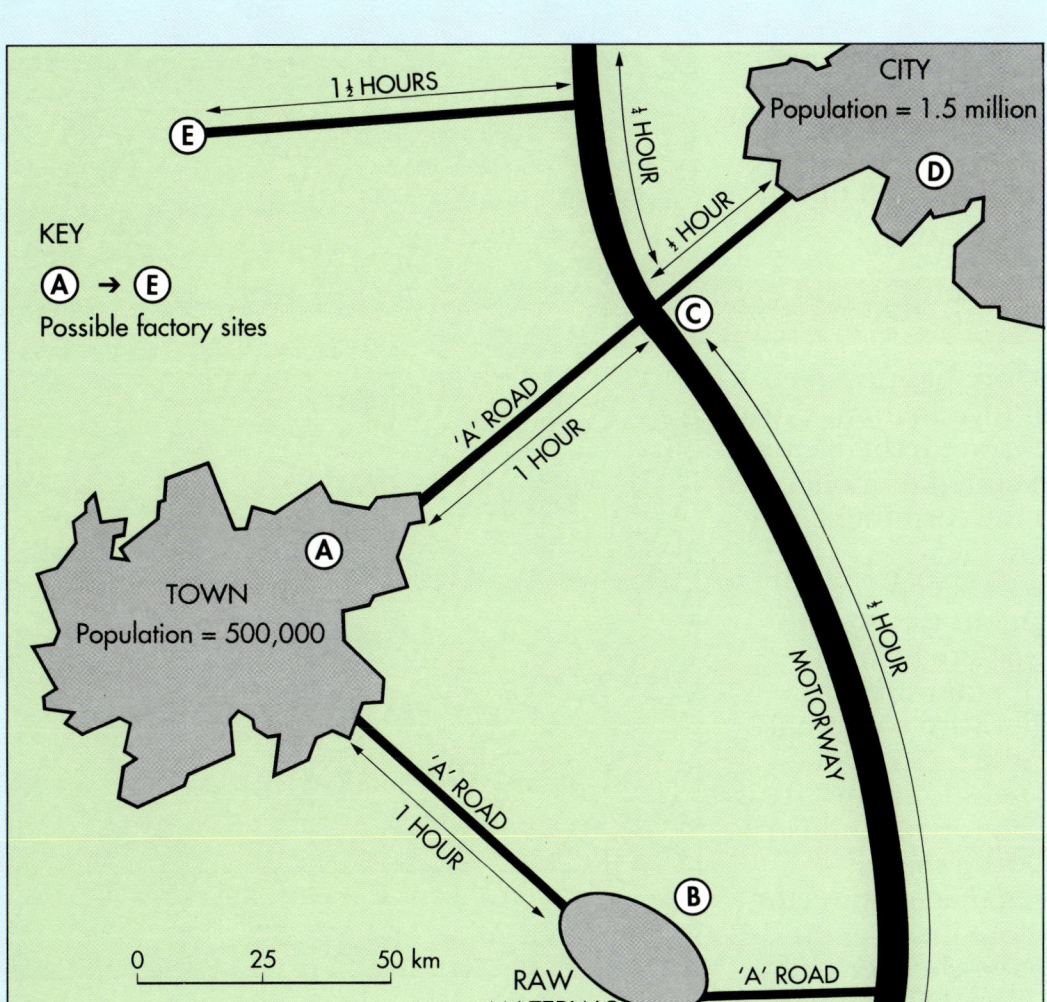

Key

A – E Possible factory sites

A large town location

B Raw material location

C Motorway location

D City or conurbation location

E Out – of – town location

	Score
LABOUR Travel distance of factory from city of 100,000 people:	
½ hour	3
½ –1 hour	2
More than 1 hour	0
RAW MATERIALS	
Available directly on site	4
1 hour from factory (or less)	3
More than 1 hour from factory	0
TRANSPORT	
Motorway ½ hour or less from factory	2
Motorway ½ hour to 1 hour from factory	1
Motorway more than 1 hour from factory	0
MARKET	
1 million people within 1 hour	7
500,000 people within 1 hour	5
POWER Equally available at every site	

	Labour	Raw materials	Transport	Market	Total
Site A					
Site B					
Site C					
Site D					
Site E					

3·14 *Finding a place for manufacturing industry.*

The factory in this exercise is imaginary or *hypothetical*. But the activity gives you an idea about what it is like to make a decision about locating a factory. It does not consider all the important points that would be taken into account in the real process.

Where should the Hammill factory go?

The Hammill furniture factory has occupied the same site in the city of Birmingham since 1923. The buildings that make up the factory are now very old, and not suited to modern industry. The site, next to the Birmingham & Worcester Canal, is far from ideal.

The owner of the factory has been concerned for some time about the cost of running the business on this site near the middle of the city. The rent and transport costs are very high, and there are continual problems concerning transport, especially delays caused by traffic congestion. The city is not suited to the needs of a growing industry: access is poor and there is little space.

Ms Jane Carter, the new managing director of the company, has therefore made enquiries about moving the factory to a cheaper and more pleasant site. She will make the final overall decision, but she takes advice from the three members of her management team.

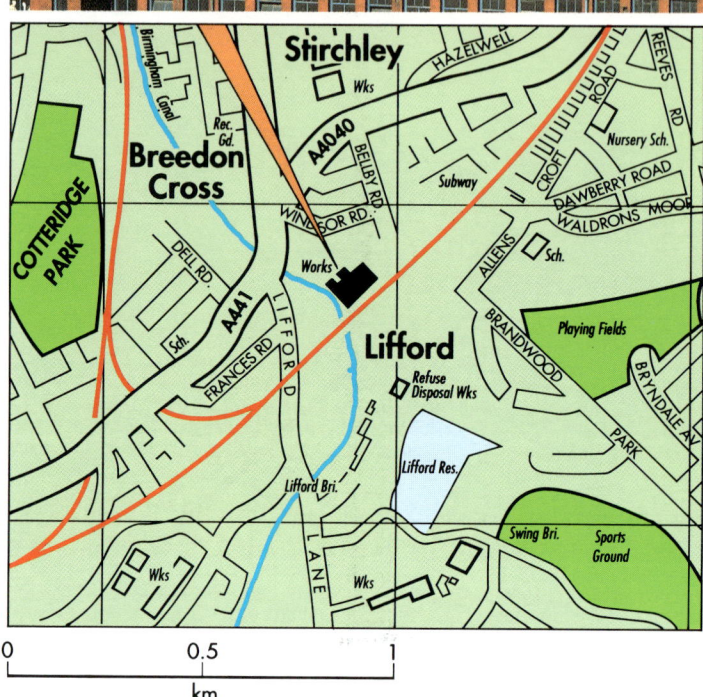

3·15 *The present site of the Hammill furniture factory.*

Ms Carter

Mr Asiz is the *financial director*. He has to work out how much money will be gained or lost by a move to a new site.

Ms Kaur is the *personnel officer* whose job it is to ensure that all staff requirements are met and that the employees are happy with the decision to move.

Mr Alexander is the *sales manager* who must decide whether sales will be increased by a change of location.

54

Ms Carter decided to hold a meeting with the management team, at which each member had to supply a brief report on the advantages and disadvantages of the two sites, at Alvechurch and Redditch, that have been suggested for the factory to move to. The decision to look at these sites followed a previous meeting when a list of ten sites was narrowed down to just two. Each of the reports below was brought to the meeting and read out.

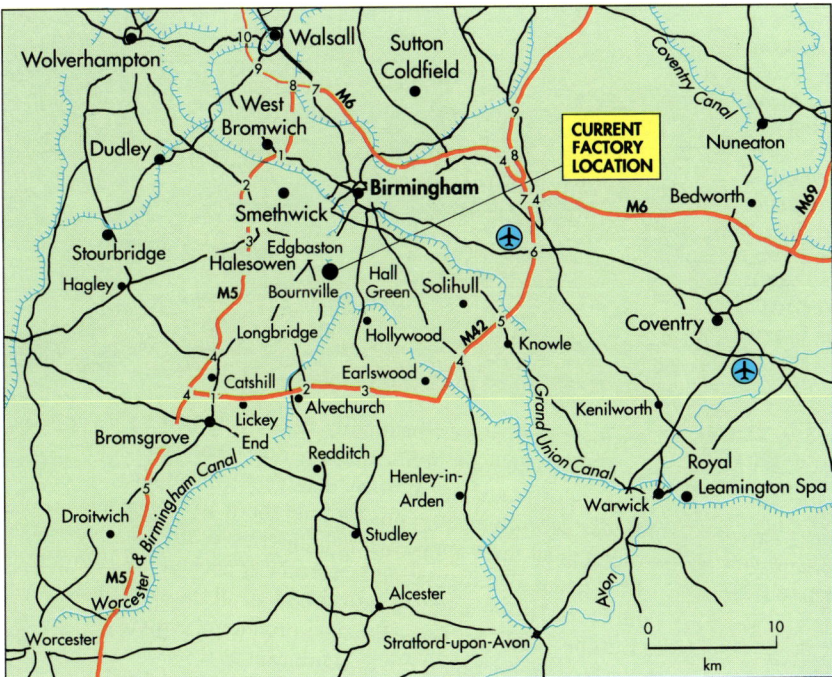

REPORT FROM : Finance Dept.
Mr. A. Asiz

ALVECHURCH
The Alvechurch site is extremely pleasant but rather expensive for renting a factory unit because it is in an area of very high-quality residential housing. Land prices are generally high owing to the closeness to Birmingham and the easy access to the M42. Planning permission for extension of a factory unit may be difficult to obtain. If employees were to move to this location they would face high housing costs. Transport costs would generally be lower than our current site.

REDDITCH
The Redditch site is in an area of recently built housing (at the end of a housing estate) and is purpose-built for a factory. Access is easy and rents are lower than the Alvechurch site and about the same as our current site. Land prices are not too high and planning permission for a factory extension would be easy to obtain. If employees were to move to this location they would face reasonable housing costs. Transport costs would be slightly lower than our current site.

REPORT FROM : Sales Dept.
Mr. J. Alexander

ALVECHURCH
A relocation in Alvechurch would not change our sales of furniture dramatically. We would still have sales in Birmingham itself and may pick up some more trade from Alvechurch and surrounding settlements.

REDDITCH
Redditch would provide a new, and possibly expanding, local market which should replace the sales lost by moving further from Birmingham. The possibilities of building up markets to the west in Droitwich, to the south in Stratford upon Avon, to the east in Warwick and Leamington Spa, and to the north-west in Bromsgrove, appear good.

REPORT FROM : Personnel Dept.
Ms. E. Kaur

ALVECHURCH
Alvechurch is an extremely attractive location for our managerial staff but is considered out of the price range by many of our office and warehouse staff.
Being close to Birmingham, 60% of the workforce said they would be happy to commute daily.
Local staff may be difficult to find for factory work.

REDDITCH
Redditch is reasonably attractive to managerial staff, although most would not live in Redditch if the factory moved there.
Being further away from the current homes of most of our workforce in Birmingham only 45% said they would commute, although 7% said they would try to move house to Redditch.
Local replacement staff would be easy to find.

3·16 *Alvechurch, or Redditch?*

Using all the evidence supplied on pages 54–55, make a decision on whether Hammills should move to Redditch or to Alvechurch. Give detailed reasons for your decision.

Sheffield, a city of change

To many people in Britain, the city of Sheffield is linked with steelmaking. It was once the major steelmaking city in Britain, and most of those who worked in manufacturing in the Sheffield region worked with steel.

Steel

The iron and steel industry has developed round Sheffield and was based originally on local iron ore and charcoal. The iron ore was found near the surface in the Coal Measures and is now exhausted. The region had other natural advantages. The deep valleys and fast flowing streams provided water power to work furnace bellows and to drive powered hammers. Local sandstone (ganister) could be used for lining the furnaces and the grit for grindstones. There was nearby limestone for flux and later the good quality coking coal of the south Yorkshire field was an important factor.

Sheffield, which lies at the confluence of the Don and Sheaf as they leave the Pennines, has a population of 495,240. In spite of the disadvantage of its distance from the coast and the absence of local ore, it is still a leading producer of steel and has built up a reputation for special high grade steels, including alloy steels. Large quantities of pig iron are now brought in from other parts of Britain and from Spain and Sweden (these last two sources provide the high quality iron needed for special steels). A considerable use is made of scrap.

There are two main branches of the industry:—

(1) A light steel industry producing machinery and cutlery. This developed on higher ground above the Don, including the Sheaf valley. The cutlery industry can be traced back to the fourteenth century and still retains some features of a craft industry.

(2) A heavy steel industry. This developed nearer to the navigable Don and continues here today. It is also found farther down the Don at *Rotherham*, where heavy steel castings and cast iron are produced, and at *Doncaster*.

From F.R. Dobson & H.E. Virgo, ' Physical and Human Geography of the British Isles', Hodder & Stoughton Ltd, 1965

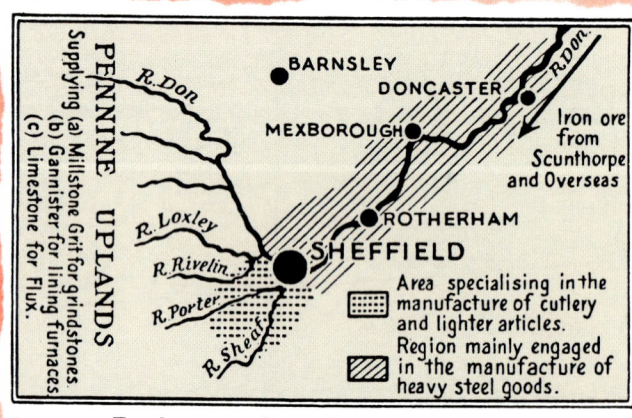

THE IRON AND STEEL DISTRICT OF SOUTH YORKSHIRE.

From D. M. Preece & H. R. B. Wood, 'Modern Geography', Book 2, University Tutorial Press, publ. 1938, reprinted 1962

3·17 *Extracts from 1960s geography textbooks.*

Sheffield is no longer the mighty steel city it once was. Most people who work in Sheffield are now employed in the service industry and not in heavy manufacturing at all. Indeed, steelworking accounts for only about 10% of people working in all the manufacturing industries locally.

Sheffield built a reputation as a steel city back in the 19th century. In 1870 it supplied almost 66% of all the steel made in Europe. It still is a specialist steel-producing centre, but its scale of production is not as significant as it was.

The 1980s brought massive changes. As well as huge job losses in the city, large areas of Sheffield had been left scarred by derelict and polluted industrial land. Since the 1960s Sheffield City Council have worked hard to improve derelict land and to clear slum terraces. Now, after a period in the mid-1980s when the valley was occupied by almost 33% of vacant and waste land, the future looks more promising.

3·18 *Sheffield in the 1990s.*

Study the items on this page.

1 List the reasons why Sheffield was an important 'steel city' in the 19th century.

2 Take three of these reasons and explain why they are no longer so important.

3 What effect would the large-scale closure of steelworks have on Sheffield?

The British Isles: a pattern of industry

The map below shows the main centres of manufacturing industry in the UK. The pattern of these industrial centres is not accidental.

Investigation

Work in small groups.

1 Examine the map, and select one type of industry to investigate in detail.

2 For your chosen industry, suggest two or three reasons why it is located in the places indicated on the map.

3 What information do you need in order to test whether your suggestions are correct?

4 Use an atlas to test your reasons. Your teacher may supply you with additional sources of information.

5 Each group should report their findings to the rest of the class.

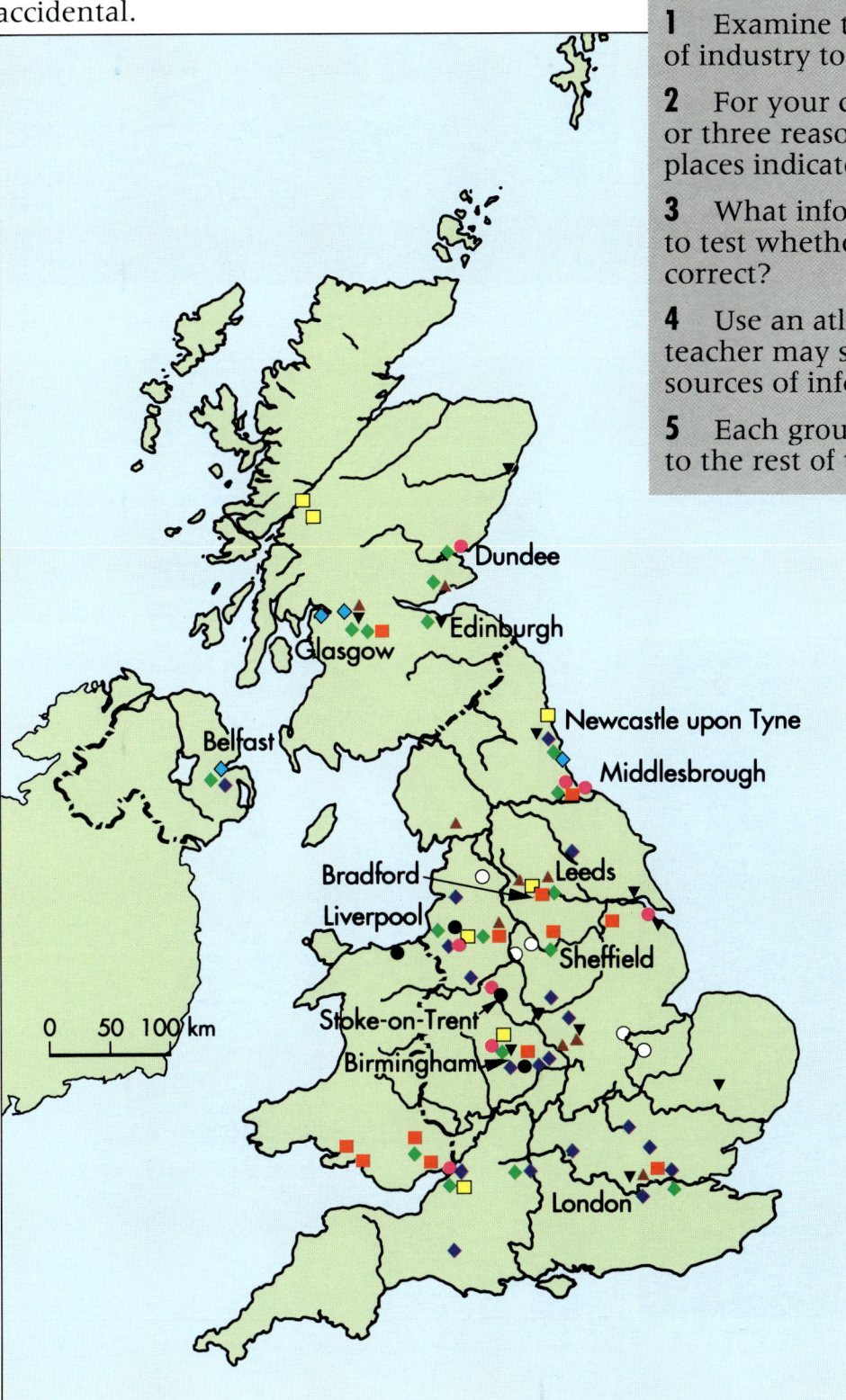

■ Metal processing
□ Non-ferrous metal processing
♦ Engineering
♦ Vehicles (road, rail and air)
♦ Shipbuilding
● Chemicals and rubber
● Glass and pottery
○ Building materials
▲ Textiles and clothing
▼ Food and drink

3·19 *Pattern of industry in the UK.*

▷ Where are the service industries in Britain?

Most of the jobs that people do in Britain today are in the service industries. These include services to *people*, such as shops, banks and libraries, and services to *industries*, such as accountancy, law, banking and insurance.

The way in which such services are supplied, and the way in which people use them, is always changing.

Offices

Services to industry are mostly organised from offices. Offices are also occupied by government services, civil servants, health, education: there is a huge demand for office space.

Where is the best place to put offices? The pictures on this page are an answer to this question.

3·21 *Offices need to be accessible because thousands of people need to commute to their place of work easily.*

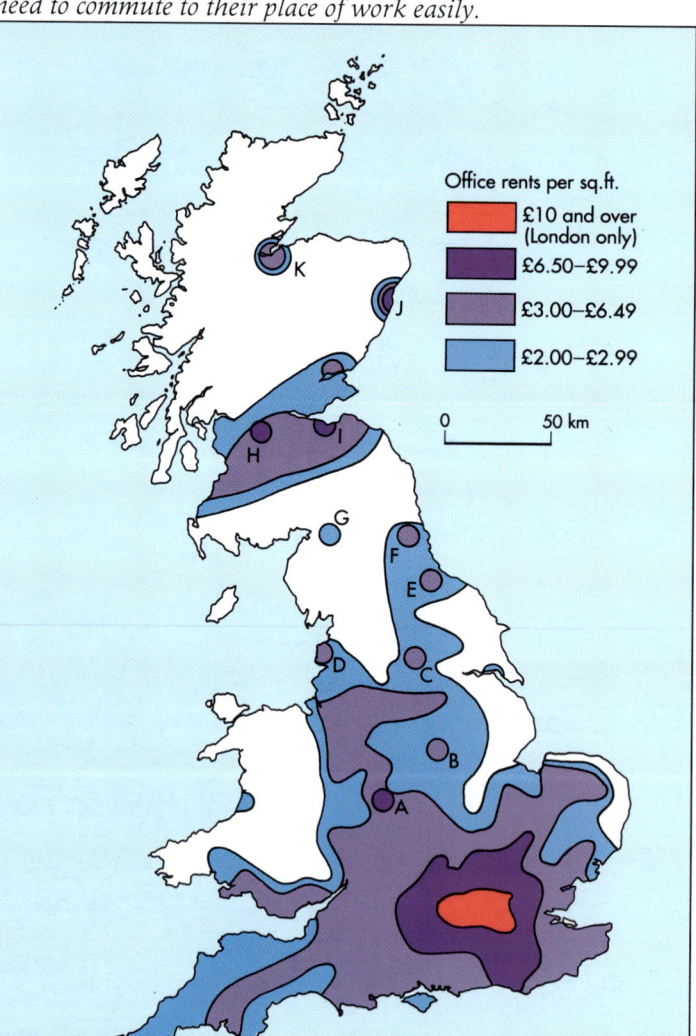

Office rents per sq.ft.

- £10 and over (London only)
- £6.50–£9.99
- £3.00–£6.49
- £2.00–£2.99

0   50 km

3·20 *The business districts of large cities are congested places, with thousands of tertiary sector jobs in high-rise office buildings. London is one of the world's major financial centres.*

Sources: Hillier Parker May & Rowden, reproduced in *Geography* Vol.75 Pt 4, October 1990.

3·22 *Office rents in Britain.*

Offices are very expensive to rent. The cost can be 200 or 300 times the cost of a house or flat of the same size. London is often considered to be the best location for a head office, but rents there are now so high that some firms, and the government, are looking at other locations. Many employees have to commute into offices from the suburbs and beyond, and many of them would prefer to be in a cheaper, regional location too.

Read the passage on this page, which was written by Mrs Sophie Williams, the head of an accountancy firm, about their decision to set up the head office away from London, in the city of Newcastle.

We decided to move from London to the North East three years ago, and have never regretted the move. Quite a few of the office staff naturally did not want to move with us, but the majority of the Board of Directors were willing to move from the South East, and we have recruited some excellent new people here in Newcastle.

Newcastle city centre has the same advantages that other large city centres offer – there are many other offices here, so 'face-to-face' contacts and meetings are easy to organise. People can get to work here much more easily than we could in London, as the centre is not so congested.

We still have the prestige of a major city address and the company image has not suffered by the move, which is important. Our clients can get to us easily by motorway, rail or air, and Newcastle has some very good hotels if our clients have to stay overnight. The city also has a wide range of other services such as restaurants and entertainments.

If I had to list the reasons why we felt the move of Head Office was important, they would be these. Firstly, the rent for office space is much cheaper than in London. Secondly, the Newcastle site is much larger and offers us opportunities for expansion. Thirdly, we have quite a lot of business with the North East and the Midlands – and a change of location helps that. Lastly, the staff wages are lower here, as we don't have to pay a 'London allowance' of £2,000 per year to each of our staff.

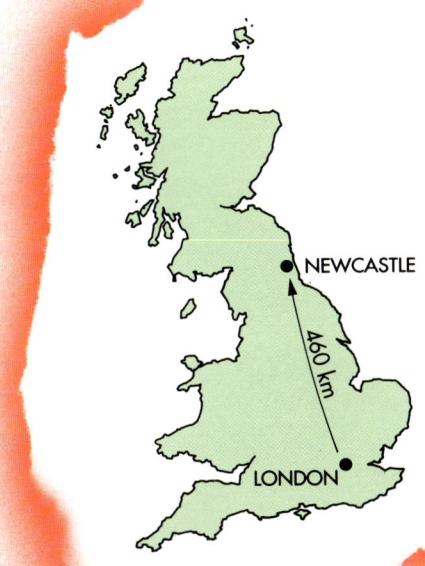

3·23 *Moving from London to Newcastle.*

The passage above explains why Mrs Williams' accountancy firm moved from London to Newcastle. The map on page 58 shows the pattern of office rents in Britain.

1 In what way is the map *evidence* of what Mrs Williams says?

2 The pattern on the map is like a contour map. Several particular locations are indicated by the letters A–K. Using an atlas to help you, name each of these places.

3 Choose one area where office rents are low.

a Name the area.
b Why are the office rents low in this area?

4 Even though offices like that of Mrs Williams have moved away from London, London and the South East of England are still very attractive to office industries.
a Read through this section and make a list of reasons why London is still an important office centre.
b Write a second list of reasons why you think London is the office centre of Britain.

Use your answers to (a) and (b) as the basis for a class discussion.

Shops

Shops provide services to people. Like offices, shopping is traditionally a town and city centre industry. Shops must be near to their customers. They must be accessible, just as offices need to be accessible to the people who work in them.

Car ownership gives people *mobility*, so that they can get around more easily. But city centres are not the most accessible places. So Britain is becoming a society of out-of-town shoppers. Most households now own at least one car, which is needed if a family is to benefit from out-of-town shopping. If you are too old, too young or too poor to own a car, this part of society is closed to you, unless there are public transport connections.

Some very large out-of-town shopping centres now serve people in whole regions, not just the local town or city. If you have the chance to stand in the MetroCentre near Newcastle for a few minutes, it is possible to hear shoppers speaking Danish and Norwegian!

The MetroCentre was built on a derelict industrial site. It is the largest shopping centre in Britain, and the fifth largest in Europe, offering shoppers not only excellent shopping facilities, but also a 'leisure experience' – it has a 10-screen cinema, restaurants, amusements and crèche facilities.

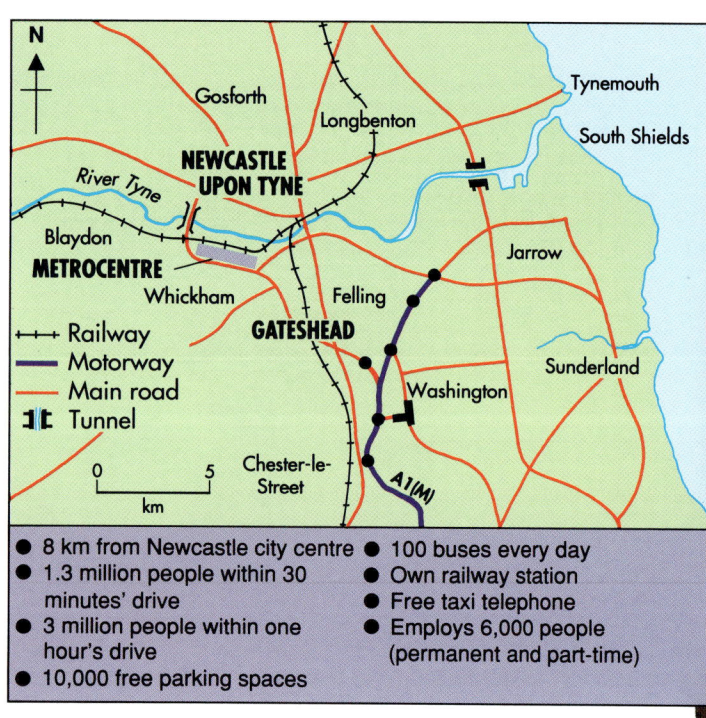

- 8 km from Newcastle city centre
- 1.3 million people within 30 minutes' drive
- 3 million people within one hour's drive
- 10,000 free parking spaces
- 100 buses every day
- Own railway station
- Free taxi telephone
- Employs 6,000 people (permanent and part-time)

3·24   *The MetroCentre in Newcastle.*

a From the air.

b Inside the MetroCentre.

The two maps below show that the MetroCentre is a truly regional shopping centre. Its shops and services have a much greater variety than those in Washington, and so people are prepared to travel from much further afield.

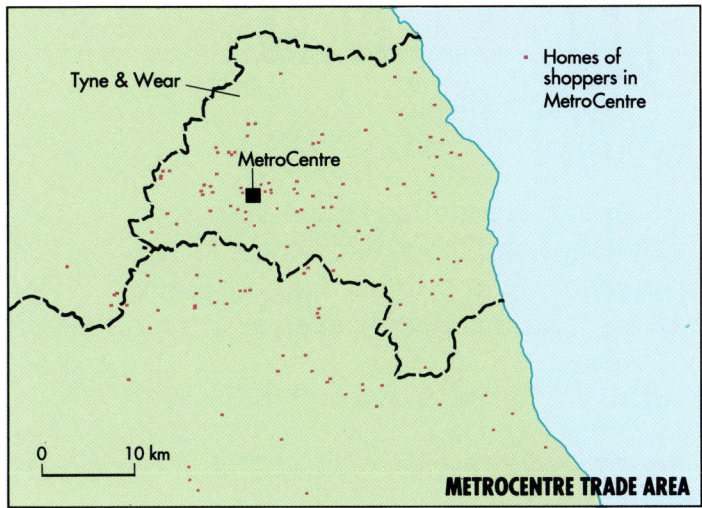

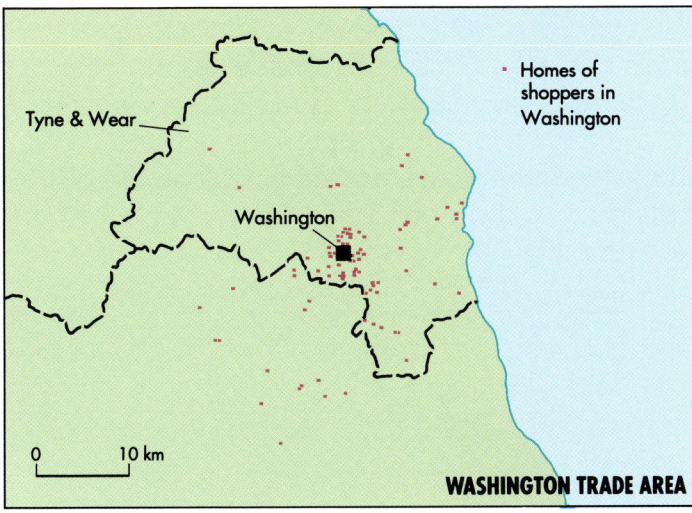

Look at the map below, which shows regional shopping centres proposed for different parts of the UK.

1 Describe the pattern on the map.

2 Give at least two reasons to explain the pattern on the map.

3 Suggest why London has relatively few very large centres. (*Hint*: Look back to the map on page 58.)

4 A huge centre is planned for Birmingham. Suggest why this is a favoured location for such a development. (*Hint*: Look at an atlas map of the area.)

Sources: OXIRM Survey; 'The MetroCentre: a new type of shopping', in A. Treadgold and E. Howard, *Geography Review* Vol. 2 No. 4, 1989.

3·25 *These two maps show that people travelling to the MetroCentre come from a much wider area than those who use the shops and services of nearby Washington.*

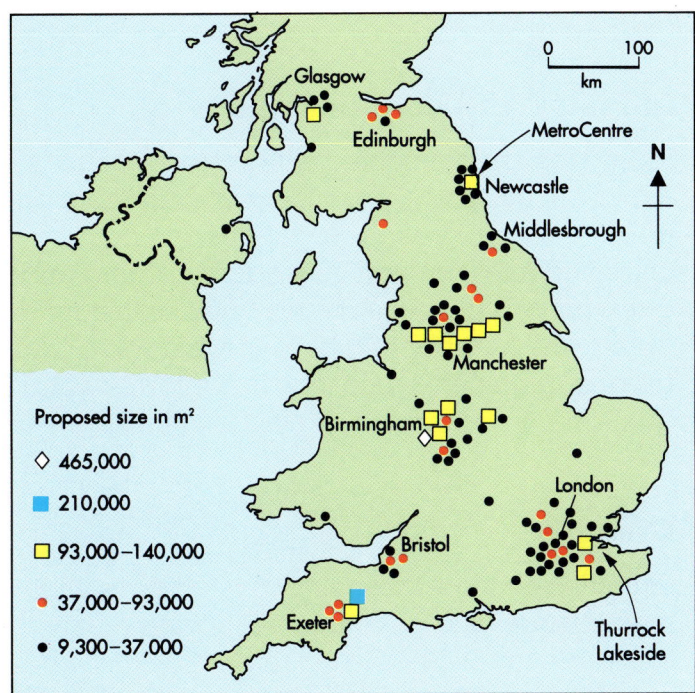

3·26 *Proposed new regional shopping centres in the UK.*

KEY ▶ POINTS KEY ! IDEAS

Location of industry

Societies need industries. Industries produce the goods and services we need, and provide jobs. Geographers have a special interest in where these industries grow and develop. Patterns of industry can be explained by looking for reasons why people locate industries where they do.

Industry and economic activity

Industries are connected with the economic activity of a society, producing and distributing goods and services. Occupations can be classified into primary, secondary, tertiary and quaternary types. In a society such as that in Britain, the number of jobs in the first two groups has fallen, while jobs in the tertiary and quaternary sectors are growing in number.

Change

Societies change. It is not always obvious why they change, and change is not always desirable for all people.

Industry is one part of our society that has changed a great deal. For example, many factories have moved. By finding reasons for this, we can understand more about why societies change.

When there are changes like this, ordinary people sometimes find it hard to adjust to the new pattern.

Mobility

Society is more mobile than ever before. Telephones and fax machines mean that *information* is very mobile indeed. Some experts think that people such as business men and women will not have to meet so often in the future. Will this mean the end of the city?

Accessibility

Most modern industries, especially service industries, need to be accessible. As mobility has increased, so has accessibility – especially in out-of-town locations. But many large shopping centres are not accessible to the old, the young or the poor, because these people lack mobility.

Making Connections

UNIT 4

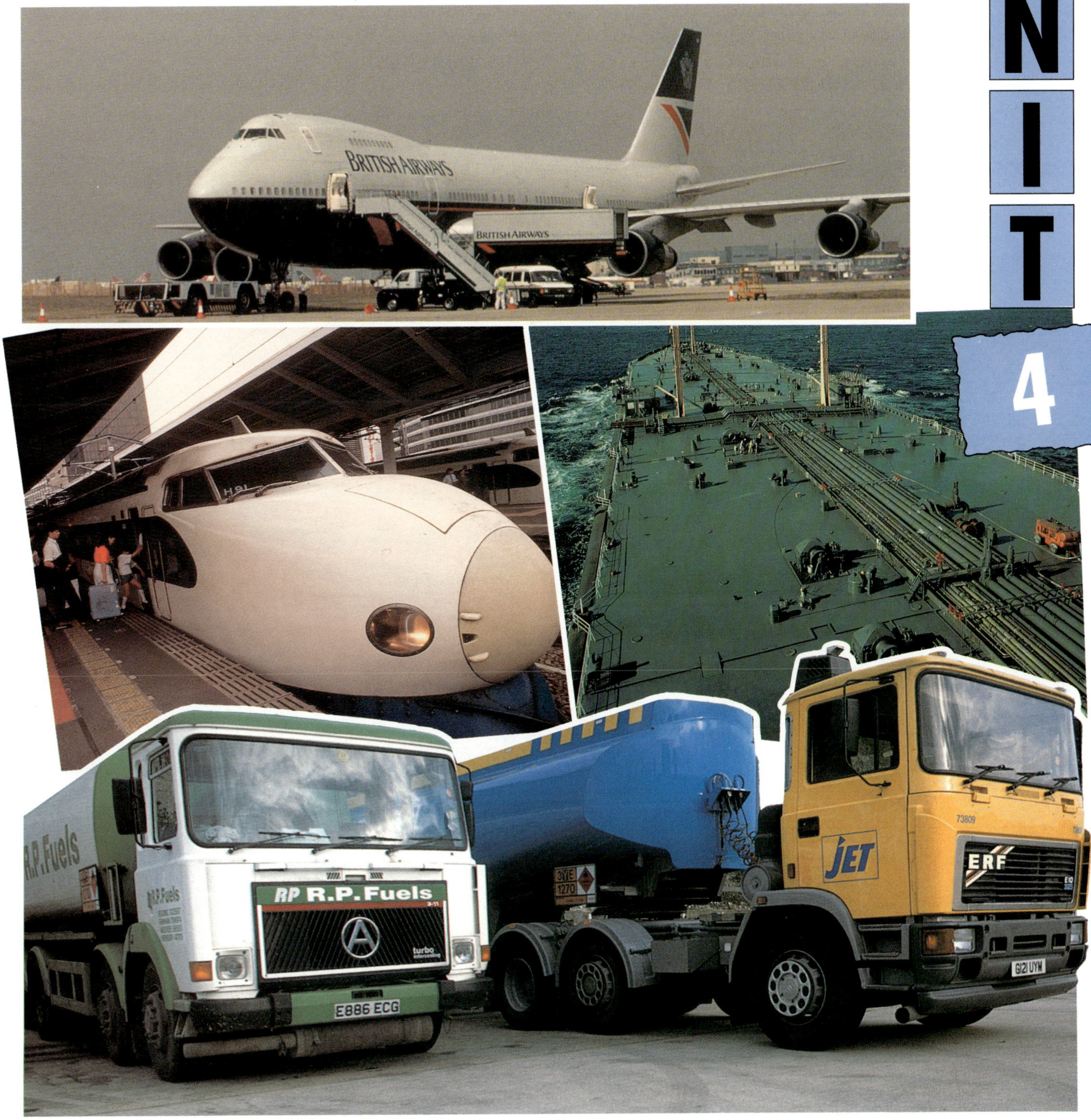

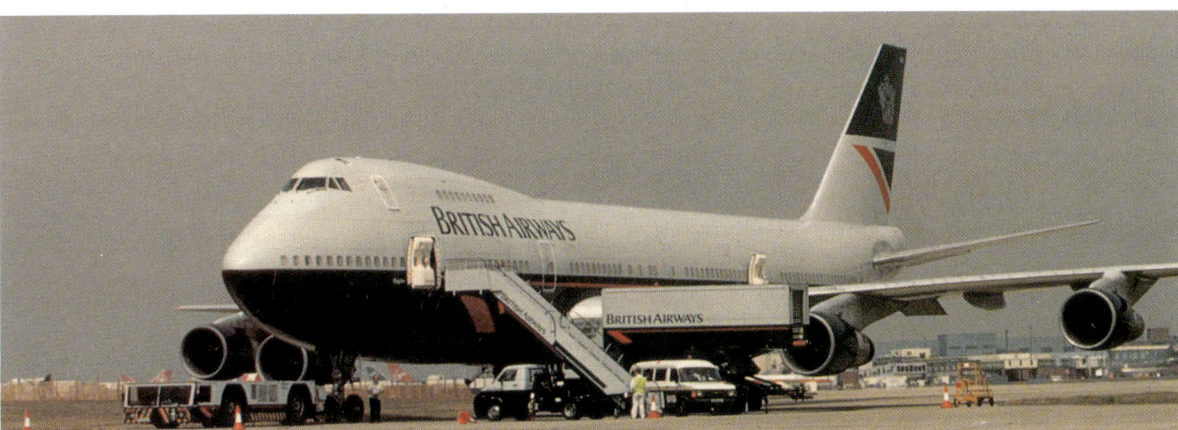

4·1 *Each of these pictures shows a different mode of transport. Transport plays a big part in making societies work.*

▶ *In what ways are societies* ***dependent*** *on transport?*

Setting the scene

Transport is the way in which people, animals and goods are moved from one place to another. Transport links are like 'arteries' carrying the 'life blood' of a nation. When people travel they are *passengers*. When goods are moved they are called *freight* or *cargo*.

Transport affects the ways in which societies work, and societies decide what transport to provide. Transport influences many aspects of

our lives. Think of how transport affects your own life. How do you get to school, go shopping, travel on your holidays? In the UK, transport uses up 17% of the average household income.

Everyone who lives in a town or city is close to a transport link of some kind. How many different types of transport are shown here?

4·2   *London's transport networks.*

With all this choice you might expect the transport system to run smoothly. But does it? What problems have you come across connected with transport? Do you agree with the views of these 13-year-olds who live in London?

> The tube and buses are so crowded.

> It's too expensive.

> The fumes are really bad.

> There aren't enough cycle lanes.

> It takes ages to get anywhere - the traffic is so slow.

> My bus is often late.

Many of these problems are connected with road transport. Roads are seen by many people as vital for Britain's development: we need to build more roads to get traffic moving faster, they say. But more roads generate more traffic, and more pollution. So what are the alternatives?

Railways

High speed, convenience, quality, fast and frequent services and real value for money all make InterCity an ideal choice for leisure travel.

b A freight train.

Principal route

Inverness
Aberdeen
Glasgow
Edinburgh
Middlesbrough
Scarborough
Manchester · Leeds · Hull
To Ireland → Liverpool · Grimsby
Holyhead · Crewe · King's Lynn
Shrewsbury · Norwich
Birmingham
Milford Haven · Bristol
London
Dover
Weymouth · Portsmouth · Hastings
Penzance

a A high-speed InterCity train.

4·3 *Does the railway system provide an efficient alternative?*

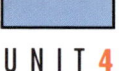

b A Channel ferry.

a A supertanker.

4·4 *Waterways are a slow but relatively cheap means of transport for both goods and people. They are essential to world trade.*

a Concorde supersonic airliner.

4·5 *In recent years air travel has become a fast method of transport for many people and for some goods, owing to reduced costs and flying times.*

b Helicopters are ideal for short journeys.

The maps opposite show 'a shrinking world'. The world isn't really shrinking – distances are the same as they have always been. But improvements in types of transport and transport networks mean that it is possible to travel from one place to another more quickly. Distances have become less important, while the cost and ease of the journey have become more important.

If a place is easy to get to, it is said to be *accessible*. Accessibility depends on

- distance
- the number of routes and links to the place
- the time
- the cost of the journey.

All different forms of transport have their good and bad points. When there is a choice, how do we decide which method to use? How do whole societies decide which forms of transport to develop?

Any decisions on transport that a society makes now will affect development in the future, at local, national and international levels. Look at news items on TV and in the newspapers to see how decisions are being made on transport developments in our society.

Transport has a big impact on how societies work and develop. To help us study this idea there are three key questions in this unit:

▷ How has transport – and society – changed?

▷ Transport choices – how do we decide?

▷ Where will transport take us?

4·6 *Is the world shrinking?*

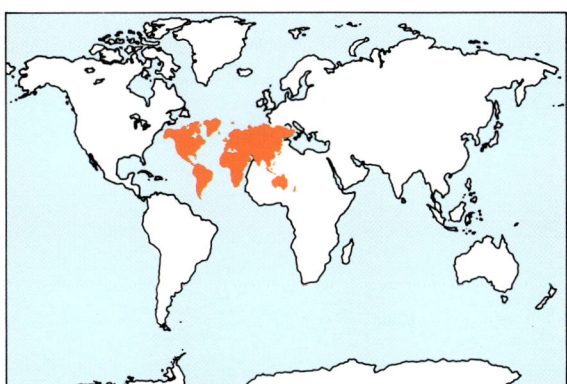

How Spain and Barbados have come closer to London

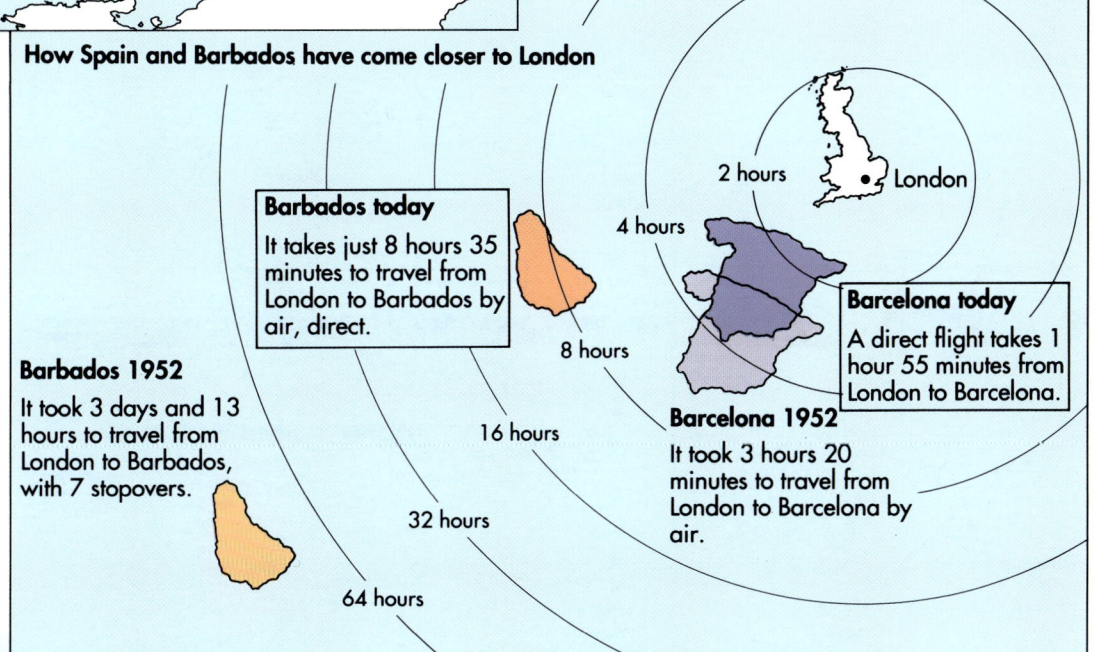

2 hours
4 hours
8 hours
16 hours
32 hours
64 hours

London

Barbados today
It takes just 8 hours 35 minutes to travel from London to Barbados by air, direct.

Barbados 1952
It took 3 days and 13 hours to travel from London to Barbados, with 7 stopovers.

Barcelona today
A direct flight takes 1 hour 55 minutes from London to Barcelona.

Barcelona 1952
It took 3 hours 20 minutes to travel from London to Barcelona by air.

Planes cause complaints

TRAINS – a step back TO THE FUTURE

New road to destroy unique Downland

Fares To Go Up - AGAIN

CHANNEL Breakthrough!

4·7 *Recent newspaper headlines.*

▷ How has transport – and society – changed?

Developments in transport have brought many benefits but they have also brought problems. The aim of this section is to find out how transport has developed in Britain, and how different sectors of society have been affected by these changes. We look at roads, railways, air and water transport.

How has Britain's road network changed?

The first roads were built in Britain by the Romans so that their army of soldiers could move around the country more easily. These roads were made of stone and cobbles, and in most places were very straight.

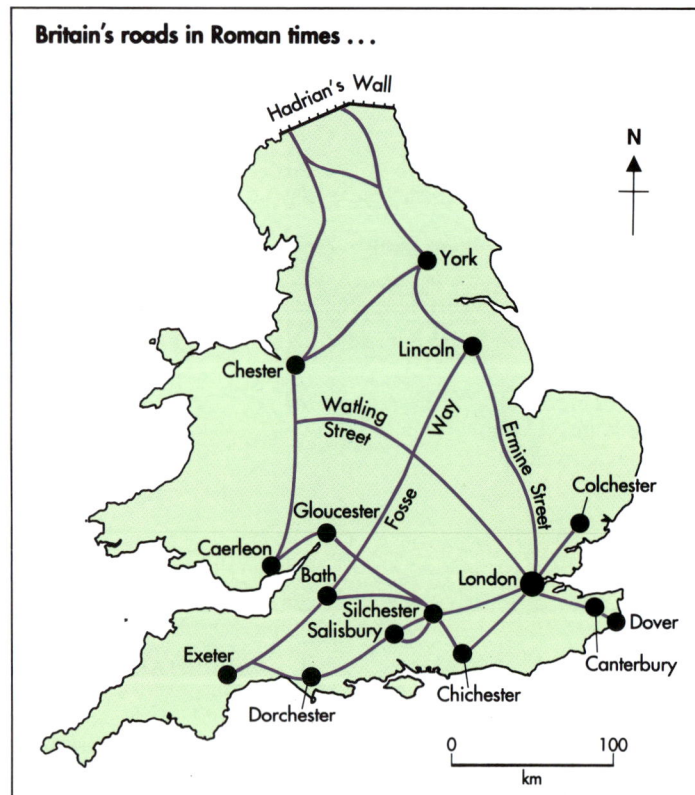

Britain's roads in Roman times ...

... and today

Since the 1940s there has been a huge increase in the number of cars and lorries on our roads. Many more roads have been built, especially by-passes to take traffic away from town and city centres.

In the 19th century, as Britain became more industrialised, people needed to move goods around faster and more easily. John McAdam developed the idea of making better roads by using pieces of broken granite, and later by binding these together with tar.

4·8 *Britain's roads over a period of 2,000 years.*

TWYFORD DOWN : LOST FOR EVER?

'Green' road cuts wood

A MOTORWAY is to be driven through a historic beauty spot as part of the Government's 'green roads' programme.

Protection bodies have fought for nearly 20 years to save picturesque Twyford Down, near Winchester, Hampshire. It includes a site of special scientific interest, an ancient monument and area of outstanding natural beauty.

But Government plans revealed last week showed that a 3.7-mile section of the M3 will cut through the area.

This means the Government has rejected a less environmentally-damaging tunnel option which would have cost about £80 million more than the estimated £36 million scheme outlined in the road report.

Full details of the Twyford Down decision were expected to be given in the House of Commons this week.

The news was fiercely criticised by both the Council for the Protection of Rural England (CPRE) and Friends of the Earth.

CPRE assistant secretary Penny Evans said: "What hope is there for the rest of the countryside when the Government cannot protect areas like this?"

She added: "We are very upset about this and about the whole report. We want the environment protected, not just a few trees and shrubs planted."

Friends of the Earth's Jeremy Vanke said: "This shows what a fallacy [falsehood] this green roads programme really is, for right in the middle of it you get plans for one of the most environmentally-damaging roads for many years."

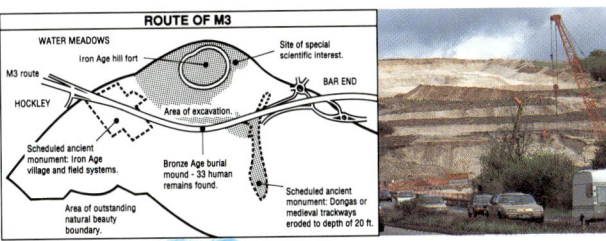

ROUTE OF M3

From the Early Times, 1–7 March 1990

4·9 *Twyford Down – lost for ever?*

Rescue diggers salvage endangered land

MECHANICAL diggers are cutting into Twyford Down, near Winchester, where the M3 is to scythe through one of Britain's most protected landscapes. But so far, the diggers belong to rescue archaeologists.

The remains of 33 bodies dating from the early Bronze Age have been found in an un-designated burial mound on the route of the London to Southampton motorway, which will slice through the down in a cutting 400 ft wide by 100 ft deep.

Opponents have asked Mr Rifkind, the Transport Secretary, to halt tenders for part of the M3 link after the European Commission said 10 days ago it would challenge the scheme because it did not comply with EC environmental law.

The decision to build the M3 through the two ancient monuments, two sites of special scientific interest – the Itchen watermeadows and the downland home of the Chalk Hill Blue butterfly – and part of the East Hampshire area of outstanding natural beauty, was taken by Mr Cecil Parkinson, when he was Transport Secretary.

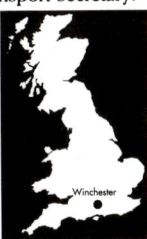

Item from the Daily Telegraph, 28 February 1990

REPORT

The government has approved the building of the road through the protected landscape of Twyford Down, to replace the congested A33. However, promises have been made to protect the area from the full impact of the road. Measures include:

- creating new open space by landscaping the A33 bypass which will no longer be used
- reducing the visual impact of the motorway with landscaping
- planting extra vegetation to keep certain animal species in the area
- carrying out an archaeological survey of the area being destroyed.

1 Compare the two road maps opposite.
a Which places had more than two road connections in Roman times?
b Which places have the most road connections today?
c Are they the same places?

2a Describe how the road network has changed.
b Why do you think the network has changed?

3 Look at the developments on Twyford Down.
a Use an atlas to locate Twyford Down.
b Describe the local area.

c What were the transport choices?
d What decisions were made? Who made these decisions?

4 You are a journalist. Write a short article (about 150–200 words) on the Twyford Down development, explaining its good and bad points. Include two short imaginary interviews. One should be *for* the road, and the other *against* it. In each case, state who the person is, his or her occupation, and where they live. Finish your article with a paragraph saying whether you think the motorway extension should have been allowed. Do you think the decision was a fair one?

How has rail transport changed?

British Rail representative Ms Briggs tells a pupil about developments in rail transport.

Jon How has the railway network changed in recent years?

Ms Briggs British Rail has spent a lot of money in the last 30 years on modernising the railways. The development of fast InterCity trains has meant that journey times to many places have been reduced dramatically in recent years.

4·10 *Today's journey times by rail from London to other parts of Britain.*

Jon Isn't the map a strange shape?

Ms Briggs It's been drawn to show the *time* taken to travel certain routes, and not the *distance* between places. The journey to some places is faster than to others, regardless of distance. Perhaps you can work out why? The diagram opposite explains why some routes take detours. The same principles work for roads too, though roads can be built on steeper slopes. In Britain, the steepest slope or gradient on a railway is less than 3%, and few have a slope of more than 1%.

Jon I see. Is that why so many areas in the country have no rail service?

Ms Briggs That's partly the reason. It costs a great deal to build a railway, and it's expensive too to keep railways running, especially the smaller 'branch' lines. The modern network is much smaller than it once was. Many cuts were made after 1963, when Dr Beeching, who was then the head of British Railways, decided to close lines that weren't making money. More and more people and goods were being transported by road, so hundreds of lines and stations were closed. The government provided money to improve other services, especially InterCity. This service now has to compete with airlines on the longer routes in the UK.

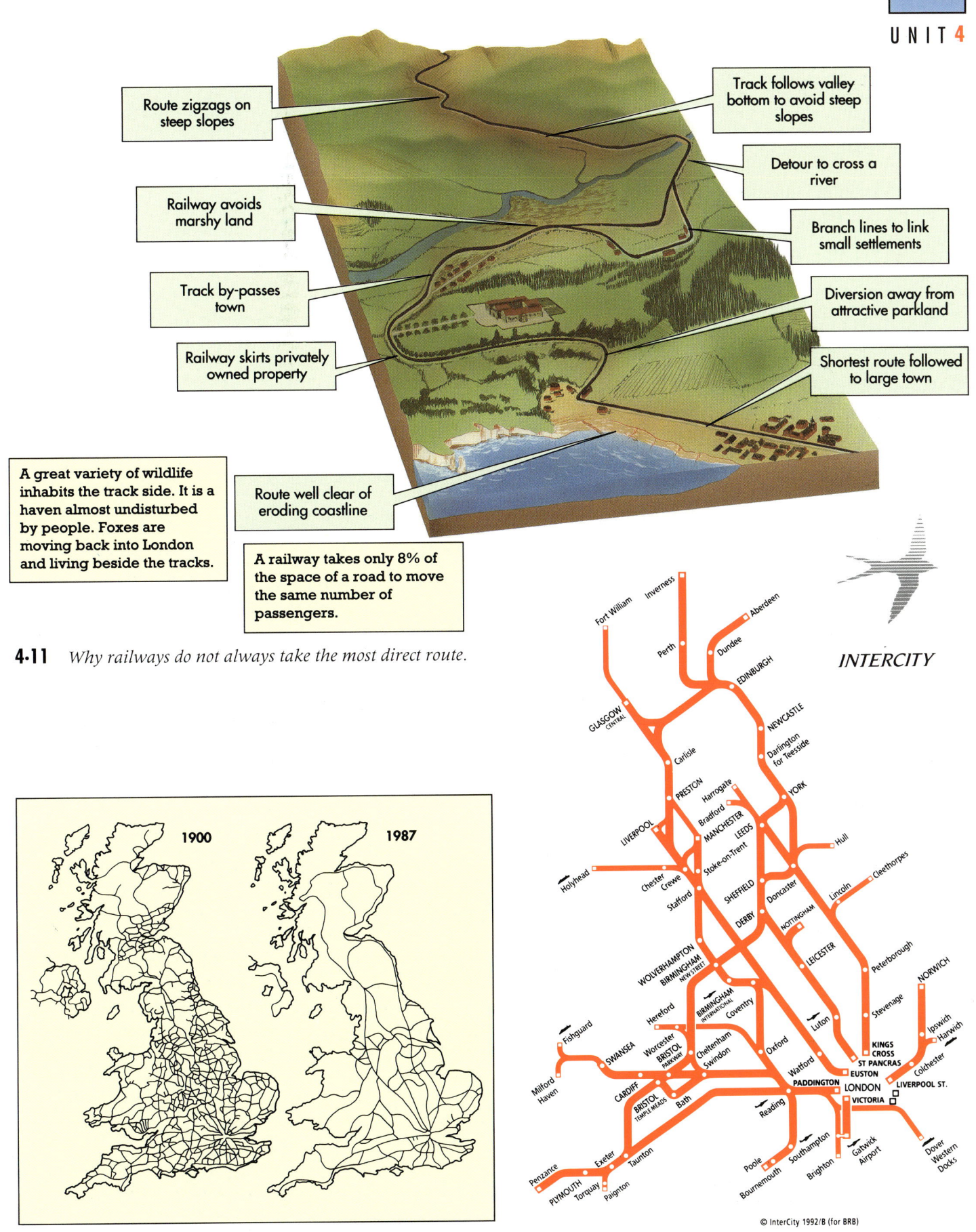

Route zigzags on steep slopes

Track follows valley bottom to avoid steep slopes

Detour to cross a river

Railway avoids marshy land

Branch lines to link small settlements

Track by-passes town

Diversion away from attractive parkland

Railway skirts privately owned property

Shortest route followed to large town

A great variety of wildlife inhabits the track side. It is a haven almost undisturbed by people. Foxes are moving back into London and living beside the tracks.

Route well clear of eroding coastline

A railway takes only 8% of the space of a road to move the same number of passengers.

4·11 *Why railways do not always take the most direct route.*

1900

1987

4·12 *The UK railway network, 1900 and 1987.*

INTERCITY

Fort William
Inverness
Aberdeen
Perth
Dundee
EDINBURGH
GLASGOW CENTRAL
NEWCASTLE
Carlisle
Darlington for Teesside
PRESTON
Harrogate
YORK
LIVERPOOL
Bradford
MANCHESTER
LEEDS
Hull
Holyhead
Chester
Stoke-on-Trent
Crewe
Lincoln
Cleethorpes
Stafford
SHEFFIELD
Doncaster
DERBY
Nottingham
WOLVERHAMPTON
LEICESTER
Peterborough
BIRMINGHAM NEW STREET
NORWICH
BIRMINGHAM INTERNATIONAL
Coventry
Hereford
Stevenage
Luton
Ipswich
Harwich
Fishguard
SWANSEA
Worcester
Cheltenham
Oxford
KINGS CROSS
ST PANCRAS
Colchester
Milford Haven
BRISTOL PARKWAY
Swindon
Watford
EUSTON
LIVERPOOL ST.
CARDIFF
BRISTOL TEMPLE MEADS
Bath
PADDINGTON
LONDON
VICTORIA
Reading
Penzance
Exeter
Taunton
Poole
Southampton
Gatwick Airport
Dover Western Docks
Torquay
PLYMOUTH
Paignton
Bournemouth
Brighton

© InterCity 1992/B (for BRB)

4·13 *InterCity network, 1992.*

What shape is Britain?

Look at the map of rail journey times from London on page 70, and the InterCity network map on page 71. One conclusion we can draw from these maps is that London is the centre of the railway network in Britain. Work in pairs for this exercise.

1 'These maps give a "London" view of Britain.' To test out how true this conclusion is, use the InterCity map to plan a railway journey between these places:

Stevenage and Luton
Norwich and Peterborough
Liverpool and Derby
Newcastle and Oxford
Cleethorpes and Birmingham

How many of these go through London?

2a Find the nearest InterCity station to where you live. Which are the fastest train journey times to the places shown in the table above, from your nearest InterCity station?

b Try to draw a 'distorted' map of Britain that shows journey times from your nearest InterCity station. Your teacher will help you with this.

Place	Journey time	Place	Journey time
Wick		Newcastle	
Inverness		Holyhead	
Fort William		Birmingham	
Oban		Norwich	
Aberdeen		Swansea	
Glasgow		Cardiff	
Edinburgh		Bristol	
Blackpool		Dover	
Manchester		Brighton	
Liverpool		Southampton	
Leeds		Plymouth	
Hull		Penzance	

Changes in air transport

Compared with road and rail, flying is a very recent form of transport. Nevertheless, airlines carry millions of passengers and large quantities of freight each year.

Heathrow, near London, is the world's leading international airport, with nearly twice as many passengers as its nearest rival, New York.

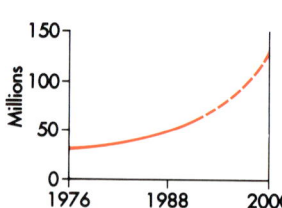
Passengers at Heathrow and Gatwick

Passengers at London's airports, 1992	
Heathrow	44,402,500
Gatwick	19,862,300
Stansted	2,314,000
Total	66,578,800

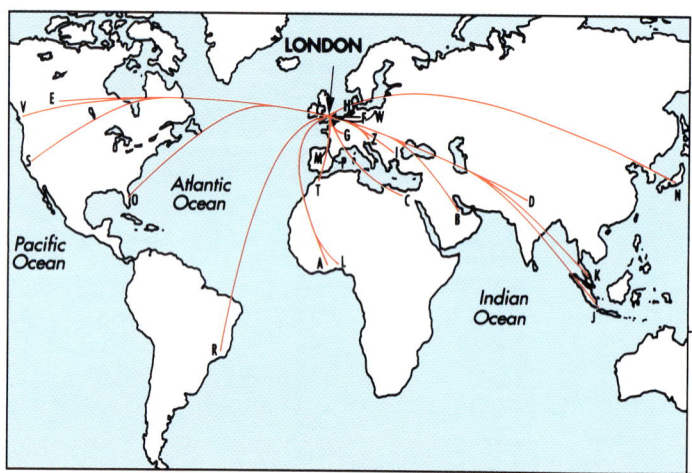

Note: Only the first location for each letter of the alphabet in the official timetable, is shown on this map.

A	Accra	G	Geneva	L	Lagos	S	San Francisco
B	Bahrain	H	Hamburg	M	Madrid	T	Tangier
C	Cairo	I	Istanbul	N	Nagoya	V	Vancouver
D	Delhi	J	Jakarta	O	Orlando	W	Warsaw
E	Edmonton	K	Kuala Lumpur	P	Palma de Mallorca	Z	Zagreb
F	Frankfurt			R	Rio de Janeiro		

4·14 *Heathrow, Gatwick and Stansted, London's three airports, handle direct, non-stop flights to more than 185 destinations worldwide. To some large or relatively close destinations there are several flights every day.*

More air travel means that the air lanes and airports become congested. Planes are more likely to be delayed, and there is a greater risk of accidents. One way of solving these problems is to build more airports, or to add extra runways to existing airports. A new airport was opened at Stansted in 1991 (see the map on page 64), and there are plans to build a fifth terminal at Heathrow.

1 What are the advantages of concentrating services at one airport rather than building new airports? What are the disadvantages?

2a How might people react to an additional runway at Heathrow? Make two lists: in the first name any people or groups who you think would support expansion, and in the second name any person or group who you think would be against another runway.

b Choose two people or groups from each list and say why you think they would be *for* or *against* the runway.

3 The map on page 72 shows the *spread* of destinations from London.
a Use an atlas to identify the countries in which these cities are located.
b Use the information on page 72, and any other relevant information you can find, to design a poster for the London Tourist Board.

Water transport: changes in Docklands

There have been big changes in the area of London known as Docklands. The following extracts are from a pupil's fieldwork notebook.

Tom English, an ex-dock worker, told me the background to recent developments.

'Today's ships are very large compared with ships of a hundred years ago. Some ports now specialise in handling one type of cargo, and are therefore more efficient. For example, Milford Haven in Wales handles oil and oil tankers. Other ports, like Harwich, Felixstowe and Tilbury, handle containers and container ships. Containers are large, strong metal cases all of the same size. They are easily loaded onto and off ships, trains and lorries.,

Tom took me around a dock on the Thames, in east London. The areas we saw looked deserted. He said the area had declined because of the success of the new container ports. Also, the old docks had little room to expand because they were surrounded by houses. These docks couldn't handle the modern ships and new methods. Then we went to a different area in Docklands where re-development is going on. Tom had strong views about the new developments.

'I was born on the Isle of Dogs, and I worked here all my life until they closed the docks in 1980. Now all the warehouses and factories we worked in have been knocked down and flats and offices built instead. The council have hardly any money to help us out—our houses are getting worse whilst we watch money being poured into these flats that we can't afford, and offices where we can't get jobs. The government said we'd benefit from the changes. This might be true eventually, but it's tough coping with all this change, especially if you're out of work and feeling ignored.'

4·15 *Extracts from a pupil's notebook.*

4·16 *A container ship.*

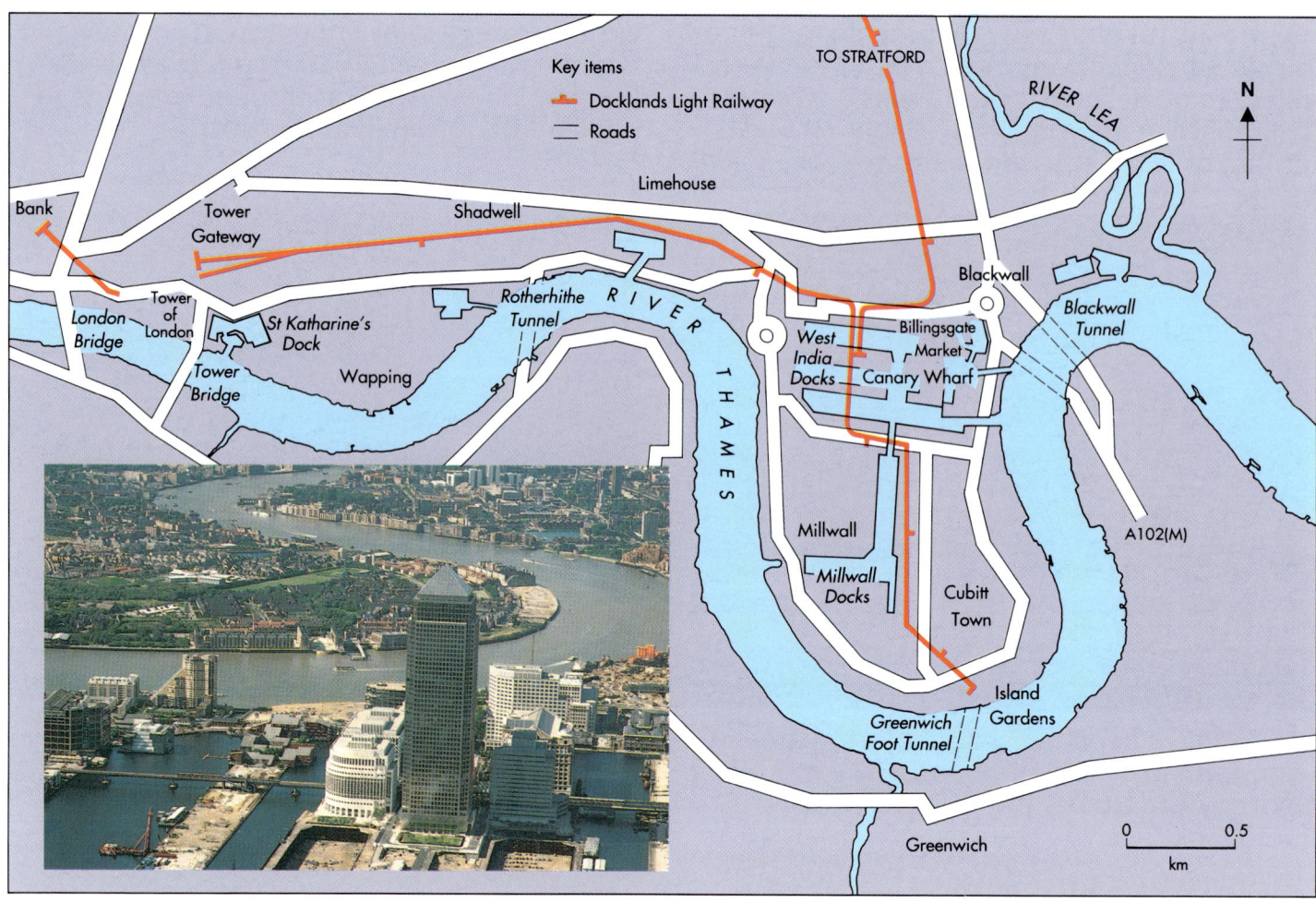

4·17 *London's Dockland development.*

1 For each of the forms of transport we have examined in this section, analyse the impact of its recent development on people. To do this, complete a table like the one below. 'Water' has been done for you.

	Water	Air	Rail	Road
Changes	*New, larger ships Old docklands close down*			
Winners	*Property developers making a profit out of redevelopment*			
Losers	*Dockland communities facing unemployment and hardship*			

2 People use transport to travel from place to place so that they can make contact, or communicate, with other people. One modern form of communication is the fax machine. It is estimated that in Britain today there are 1 million fax machines in use, and that on every working day, between 40 and 50 million pages of information are transmitted through the telephone network.

a Try to imagine how electronic machines will change transport in the future. Choose one form of transport and describe the changes you foresee.

b Who will be the 'winners' and who will be the 'losers' as a result of these changes?

▷ # Transport choices – how do we decide?

Before every journey, a decision must be made. Individuals must decide what is the best way of getting themselves or their goods from one place to another. Each method of transport has its advantages and disadvantages. The type of transport chosen depends on many things. This section shows how whole societies must make choices too.

4.18 *Which form of transport to use?*

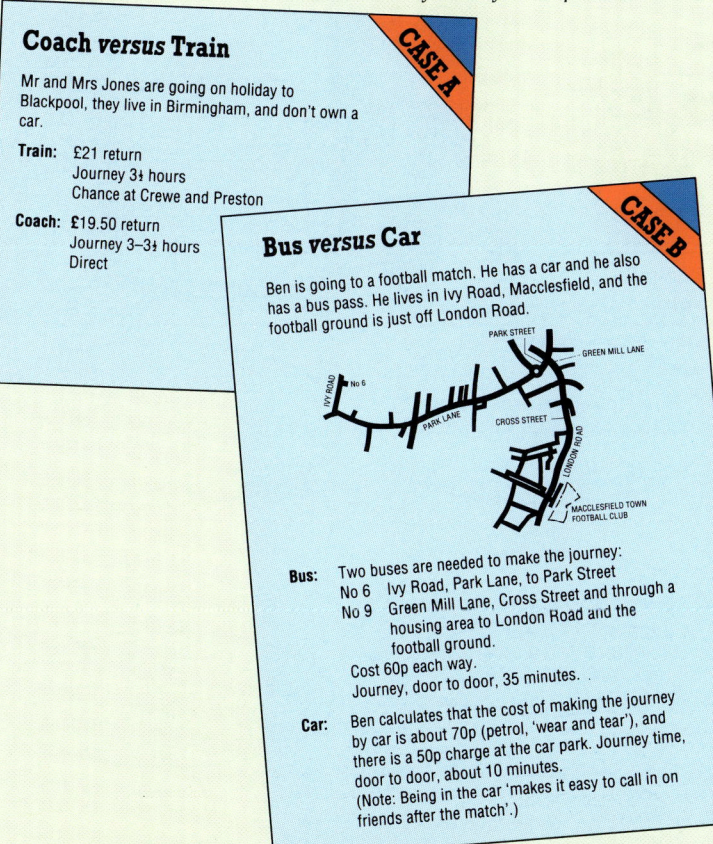

Coach *versus* Train CASE A

Mr and Mrs Jones are going on holiday to Blackpool, they live in Birmingham, and don't own a car.

Train: £21 return
Journey 3¼ hours
Chance at Crewe and Preston

Coach: £19.50 return
Journey 3–3¼ hours
Direct

Bus *versus* Car CASE B

Ben is going to a football match. He has a car and he also has a bus pass. He lives in Ivy Road, Macclesfield, and the football ground is just off London Road.

Bus: Two buses are needed to make the journey:
No 6 Ivy Road, Park Lane, to Park Street
No 9 Green Mill Lane, Cross Street and through a housing area to London Road and the football ground.
Cost 60p each way.
Journey, door to door, 35 minutes.

Car: Ben calculates that the cost of making the journey by car is about 70p (petrol, 'wear and tear'), and there is a 50p charge at the car park. Journey time, door to door, about 10 minutes.
(Note: Being in the car 'makes it easy to call in on friends after the match'.)

Plane *versus* Train CASE C

Mrs Mortimer has a two-day business conference in Edinburgh. She lives in Victoria, London. Her company will pay travel expenses.

Plane: Heathrow to Edinburgh
£160 return
Journey 1 hour

Train: Kings Cross to Edinburgh
£57 return
Journey 5 hours

Tube: Victoria to Kings Cross
¼ hour, 80p
Victoria to Heathrow
¼ hour, £1.70

Edinburgh airport to conference 30 minutes by taxi, £6.00
Edinburgh station to conference 10 minutes' walk

Van *versus* Rail CASE D

A company based in Liverpool needs to transport some of its products to Cleveland, where they will be sold.

Rail freight:
Lorry–train–lorry
1–2 days
About £250, including packing

Removal company: Lorry door to door
5 hours
£234

Intermodal transport

Red Star is a delivery service that uses more than one type of transport. This is known as intermodal transport. Parcels are collected from customers by lorry or van and then transferred to travel long-distance journeys by train. The parcels are delivered at the other end by lorry. The guaranteed same-day delivery is possible because a combination of modes of transport is used.

1 Look at each of the four case studies above. In groups, discuss each of the alternatives. Which would you choose in each case? Explain your choice.

2 Draw a flow line diagram, showing modes of transport used, for the journey of a package sent by Red Star, from Norwich to an address in Newport, South Wales.

At peak times, one bus carries the occupants of 22 cars.

How is it they're so punctual with the fare increases?

I never use trains. They're so overcrowded!

U.K. MOTORWAY

This plan still leaves room for **dozens** of houses.

Knock the next lamp on the left sideways, crack the pavement by the pub, scrape the grocer's shop, drive over the roundabout, then follow the traffic jam to London.

There are 24 million vehicles on Britain's roads.

There are 24 million vehicles on Britain's roads.

Road transport produces 19% of the UK's CO_2 and 88% of CO.

Between 1981 and 1991, rail fares went up by 51%.

New road schemes threaten 2 National Parks, 12 Areas of Outstanding Natural Beauty, and 160 SSSIs.

Car ownership may double by the year 2025.

53% of women and 22% of men do not have a driving licence.

Over 60% of all freight is carried by road. 7% is carried by rail.

A rail ticket costs 2½ times as much in Britain as in France.

In 1971, 80% of 7–8 year olds in England went to school on their own. In 1990 only 9% did.

The extra traffic expected by 2025 will take up the space of a motorway from London to Edinburgh 257 lanes wide.

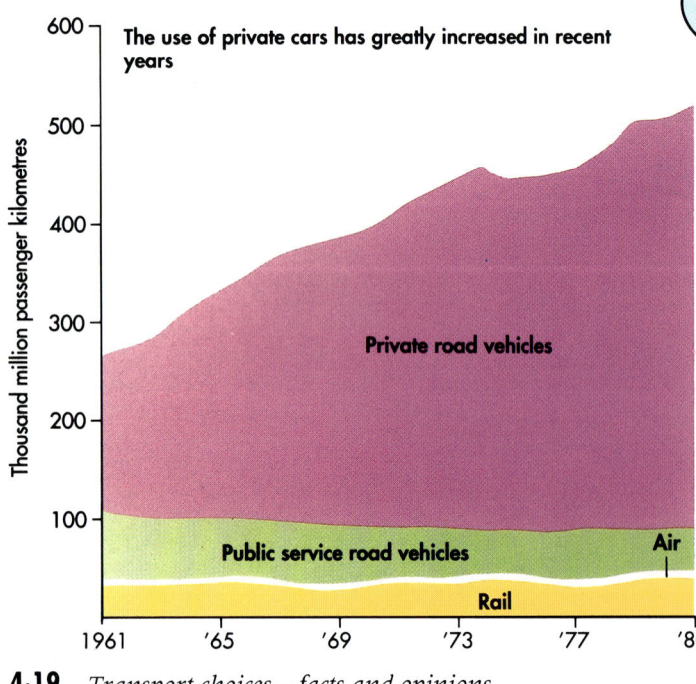

The use of private cars has greatly increased in recent years

Thousand million passenger kilometres

Private road vehicles

Public service road vehicles

Air

Rail

1961   '65   '69   '73   '77   '80

4·19 *Transport choices – facts and opinions.*

Transport choices face the whole of society as well as individuals. Examine all the facts and opinions presented on page 76.

1 In small groups, take it in turns to read out the statements and cartoons. Make sure you understand each item, including the graph.

2 Many of the statements refer to transport problems. In your group, discuss what choices these problems present to society. For example, the prediction that 'Car ownership may double by the year 2025' gives present-day society a choice:

● Do we encourage more people to own and use cars, *or*
● Do we discourage car ownership?

3 Following your discussion, write down what transport choices face Britain.

Public or private transport

There are two main ways of providing more transport. The first is to allow *private transport* to increase. This means building more roads to allow for greater numbers of cars and lorries. The second is to improve *public transport*, such as buses and trains, and to discourage private transport. However, this is not just a simple choice. The best solution is probably to do a bit of both.

Below are some of the ways of encouraging private transport. Can you think of others?

1 Think of as many ways as possible of discouraging private transport, and improving public transport. Write down your ideas, using diagrams or pictures if you wish.

2 Draw a 'link' diagram to show how public and private transport in towns affect each other. Write down the following phrases, each in a box, and add any others you feel are necessary. Fill a whole page, leaving plenty of space between the boxes. Then draw lines between the boxes to show any links between them.

Poorer-quality public transport

Increase in car use

Car ownership more popular

Fewer public transport users

Fewer public services

Higher prices

More traffic congestion

Better traffic management:
• roundabouts
• one-way systems
• traffic lights

More off-street parking

Parking 1,200 places

Urban motorways to improve traffic flow

By-passes to keep through traffic out of towns

Follow signs for by-pass

4·20 *How to encourage private transport.*

The government in Britain makes many transport choices that influence how society develops. Making choices is difficult, and usually controversial, as the newspaper articles here show.

TRANSPORT PLAN WANTS SERVICES CLOSER TO HOME

Planners trying to solve Britain's transport problems should concentrate on bringing shops and jobs close to people's homes or closer to better bus, rail and cycle services. Instead they concentrate on providing the means to make more and faster road journeys. Emphasis should be on reducing the distances travelled.

Road traffic is estimated to double over the next 30 years, and up to £20 billion is due to be spent on building new highways. It seems that current transport policies are making transport problems worse, not better.

Transport Department 'Biased Towards Roads'

THE DEPARTMENT OF TRANSPORT has become obsessed with roads, to the virtual exclusion of all other methods of transport, a report says. The result is that the roads programme continues apace, while investment in other modes of transport is restricted.

Items from national newspapers, 1992

1 Why is the Department of Transport being criticised?

2 What suggestions are made for solving some of Britain's transport problems?

3 Do you agree with these solutions? Why?

▷ Where will transport take us?

Much of the transport we use today would have seemed impossible even 30 years ago. Huge motorway systems, supersonic aircraft and high-speed trains have all been introduced, and transport continues to develop, reflecting the changing needs of our society.

One of the largest transport projects ever undertaken anywhere in the world is on Britain's doorstep. The items on this page and the next page describe this spectacular transport development.

On 1 December 1990, the United Kingdom was joined with continental Europe. The last time these two land-masses were joined was 20,000 years ago, during the Ice Age. The one-time dream became a reality when the pneumatic drill at the tunnel face roared, and the first hole appeared – a moment to change history.

From a newspaper article, 19 December 1990

Fog blankets the Channel. Britain is cut off by just 30 kilometres of sea. Holidays begin with delays. Vital export orders are lost. Now, at last, Britain is set to realise an age-old dream: a fixed transport link with the continent.

From a Eurotunnel publication

YOUR VIEWS ON THE NEWS

The Chunnel: will it all be plain sailing?

THE £7-billion Channel Tunnel is set to open in June 1993, cutting the 75-minute ferry crossing to a 30-minute journey under the sea.

Love it or hate it, Eurotunnel is a major breakthrough, which will offer holidaymakers and businessmen greater choice when travelling to Europe. We asked ET readers for their views.

Daniel Saunders, 16, of Herts: "I feel a bit sorry for people whose lives are going to be messed up because of the developments in the railway system - building a new line across beautiful land and all the extra noise pollution.

"I also feel sorry for the ferry companies. A lot of them have stakes in the tunnel so they don't lose out either way, but a lot of smaller companies are going to go bankrupt straight away.

"It's going to mean new jobs for some people, but a lot of people are going to lose jobs. I would guess more people are going to be out of work than in work.

"The idea of being under the water doesn't appeal to me that much. It's quite enjoyable to go for a cruise across the Channel."

Leonard Winning, 10, of Surrey: "I think it's a fairly good idea because it will be a

We asked ET readers what they thought of the Channel Tunnel

fast and efficient way to get from England to France, and from France to other places in Europe.

"I wouldn't go on it, I don't think. I would be a bit too afraid of water leaking! I think I would rather go on a boat."

Daniel Alter, 14, of Essex: "I think it's a very good idea because it will give us closer links to Europe. International business will become much closer.

"For business people who want to get to Paris quickly, I think it's a good idea. If someone is going on holiday and wants to cruise over there, going on a ferry would be better.

"It has taken a long time. There has been a lot of hype and I would like to see for myself whether it's such a great thing!

"A lot of people get worried when they take their first trip on a plane. This is a first trip, only this time it's under the ground. I'm sure it's quite safe. I probably wouldn't be too nervous."

Maxine Twynam, 13, of Bristol: "The Channel Tunnel is a bit stupid really, although I

think it will be cheaper. We could get rabies from France by rats that come over through the tunnel. There will be traffic jams. It has cost so much money to build and we don't really need it.

"I wouldn't go across to France that way. I would rather go by boat or plane."

Martha Reeves, 13, of Surrey: "I would give it a try, but I wouldn't like to be first. I would like to see someone else has been across it.

"Some people think it's terrible and get a bit silly about claustrophobia, even if they don't suffer from it.

"There's just as much chance of a boat sinking as the Channel Tunnel caving in. I don't know if I would feel safer in the tunnel or on a boat or plane. They all have hazards."

Steven Stark, 15, of Herts:

"I think it's a great idea. I've got no problems going under water. I think it's perfectly safe. It's going to be less dangerous on the train than the ferry. Each compartment is designed to withhold fire. If something catches fire, they carry straight on – they don't stop in the tunnel.

"I suppose it could be a terrorist target. It could be quite a mess if there was an attack.

"Three hours London to Paris is going to mean that England is a lot more into Europe, which is good for 1992."

*To some the tunnel will be a boon
And to them comes none too soon!
Sea journeys can be very tough
Especially if its really rough!
It's hardly any trippers wish
To turn green & feed the fish!!*

Excuse me, but was the fruit salad and double cream for you, sir?

From the Early Times, 6–10 August 1990

Breakthrough! The French and English join hands in the Tunnel for the first time

The Channel Tunnel is not a new idea. As long ago as 1802 a French engineer developed a scheme to link England with France with a passageway under the sea to carry horse-drawn trolleys. Napoleon Bonaparte expressed interest in the scheme but nothing came of it because his enthusiasm was not matched by that of the British Government. Digging actually began but the project was abandoned because of fears that it would expose Britain to French attack.

Work on a new stretch of tunnel began in 1974. More than 2 kilometres were dug on either side of the Channel before the British Government, alarmed by rising costs, backed out again.

But today the Government is not in a position to withdraw financial support – it has refused to invest *any* public money. As a result, Eurotunnel, the private company in charge of the project, has raised the money by selling shares in the project and by borrowing from banks around the world.

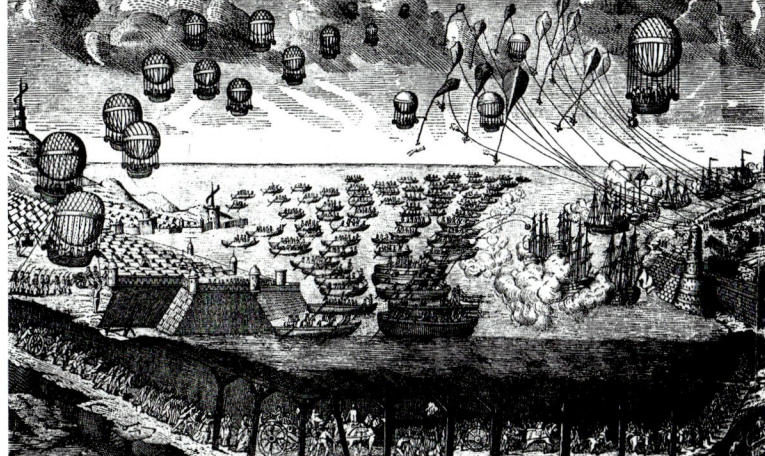

French drawing of 1803 showing early plans for a Channel Tunnel

4.21 *'Underground Movement': the Channel Tunnel.*

Work in groups for this activity.

1   On a large piece of paper, draw up a 'spider' diagram with boxes to show all that you have heard or read about the Channel Tunnel, and the links between them. Your teacher should be able to provide you with additional material.

2a   Circle all the things your group feels are *good* about the Tunnel in red, and underline all the *bad* things about the Tunnel in black.
b Are there more good or bad things? Do you think the Channel Tunnel is a good idea? Write down your answers individually, as if you were being asked for 'Your views on the news'.

Society – who benefits, who loses?

For the Channel Tunnel to be a fast alternative mode of transport to the Continent, new rail links were needed on either side of the Tunnel to run to London and Paris. There was a lot of opposition to the various proposed routes, particularly from people in the villages that lie near the routes.

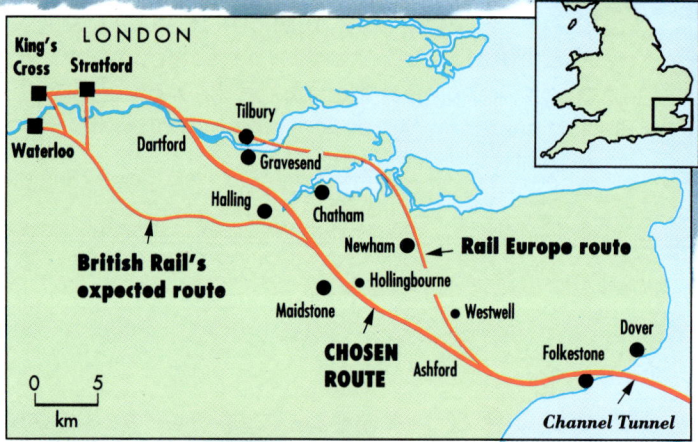

THIS GARDEN'S GOT TO BE DUG

Local people have tried to block all the direct routes from the Chunnel to London, saying "Keep Kent the garden of England".
Kent is a beautiful county. Wales, the Midlands, and the North are beautiful too, but they were ripped apart for coal and iron.
Kent cannot be allowed to block progress. A successful rail link would enrich the whole country. Gardens are lovely. But you can't live on dandelion seeds.

From a national newspaper

Outrage all down the line

By BOB McGOWAN

MPS and home-owners along the new route were outraged.
"It was bad from the start, it got worse and is now very much worse," said Dartford's Tory MP Bob Dunn.
"It's the most appalling thing," added Ashford Tory Keith Speed.
Picturesque Hollingbourne village in Kent was full of houses for sale yesterday.
Widow Mary Bignell, 73, said: "I've lived here all my life and I'm

the 200-metre corridor of the line so we won't get any compensation for the noise we will suffer."
too old to leave now but many who want to cannot because they can't sell their homes."
An entire estate of first-time buyers near Gravesend, in the path of the new link, were devastated.
"Now we have no chance of moving on to bigger homes when we can afford it," said Paul Karlsson-Willis, who lives with his wife and two children on 2,000-resident Hever Court estate.
"Most of us will also be outside

4·22   *Rival routes from the Channel Tunnel to London.*

From the Daily Express, 10 October 1991

1	What problems may the rail link bring to villages like Hollingbourne?

2	Can you think of any advantages of the link?

3	Read the press release and the statement by the CPRE below. Now draw up a list entitled, 'Measures to ensure adequate protection of the environment'.

The people of Hollingbourne consider themselves unfortunate in the decision that was made in 1991. However, other people further away are also affected by the Channel Tunnel.

Which areas of Britain will benefit most from the Tunnel, and which areas will lose out? The maps below, which show the proposals to link the Tunnel to regions away from London and the South East, may help to answer this question. The answer will become easier to see as the years pass by.

PRESS RELEASE
9 October 1991

The Department of Transport today announced that a route through the East Thames corridor has been selected for the purpose-built rail link between London and the Channel Tunnel.

The Council for the Protection of Rural England (CPRE) says this decision is more in tune with countryside protection and planning policies. However, this route is not without problems – full measures will be required to ensure adequate protection for the direct and indirect impacts of the rail link.

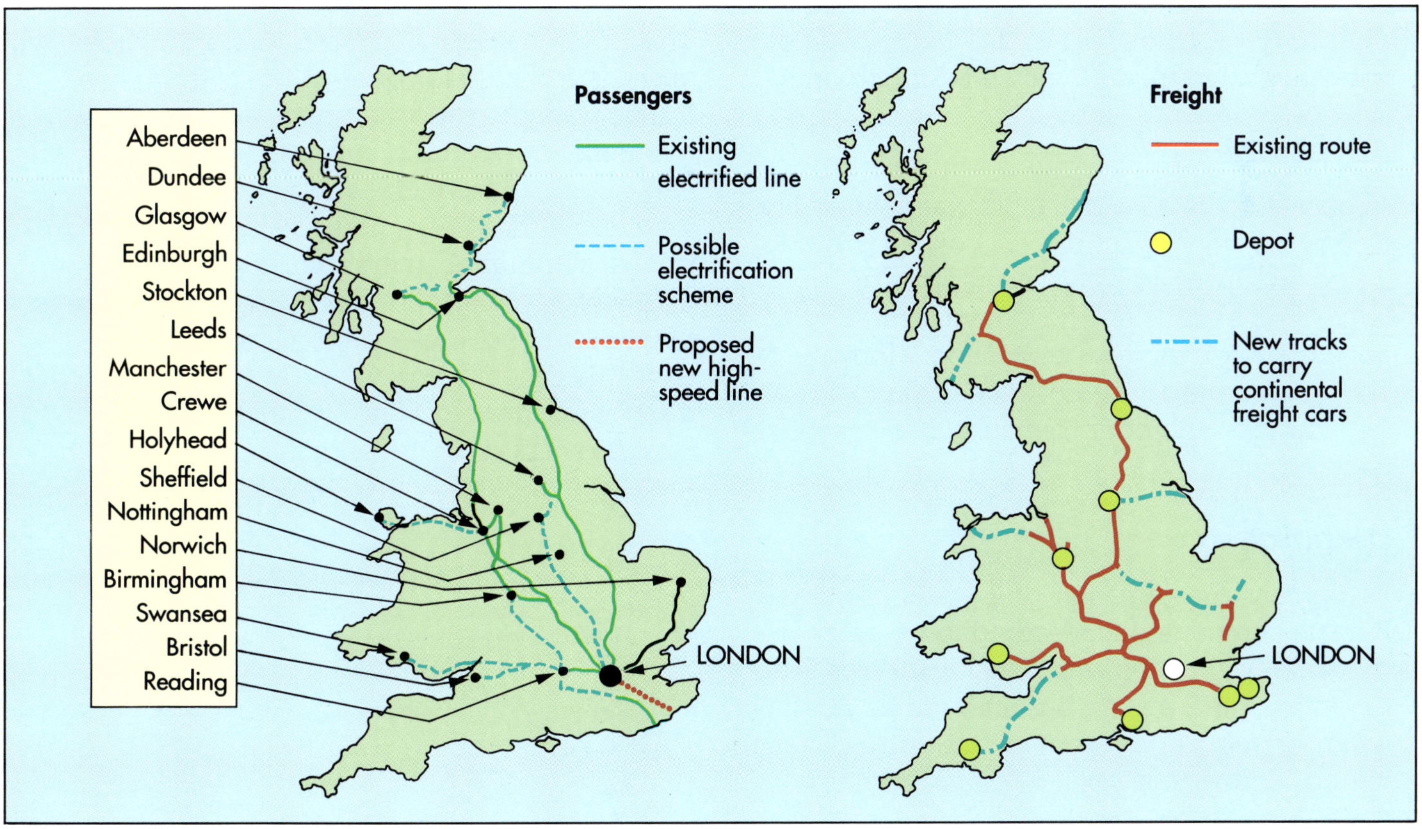

Passengers

━━━ Existing electrified line

╌╌╌ Possible electrification scheme

•••••• Proposed new high-speed line

Aberdeen
Dundee
Glasgow
Edinburgh
Stockton
Leeds
Manchester
Crewe
Holyhead
Sheffield
Nottingham
Norwich
Birmingham
Swansea
Bristol
Reading

LONDON

Freight

━━━ Existing route

⬤ Depot

╌•╌ New tracks to carry continental freight cars

LONDON

4·23	*Proposed extensions to the rail link.*

Modes of transport

There are different modes of transport linked to land, sea and air. Different modes of transport sometimes compete with each other. For instance, road transport is quite fast, flexible and expensive compared with water transport, which is slow, not flexible (it needs water!), but cheap. Often transport users are able to choose between different modes.

Mobility

Mobility is about *how* people travel from one place to another. The motor car – if you have one – gives people mobility: it takes you from door to door, when you want. Also, the car is private; it is yours and you can depend on it.

Accessibility

Accessibility is not quite the same as mobility. Accessibility is about *places*: how easy it is to get to and from a place. The centres of our large towns and cities now have less accessibility owing to congestion. Many large shopping centres are not accessible to the old, the young or the poor, because these people lack mobility.

Distance

Distance is measured in kilometres or miles. But it is sometimes misleading to measure distance only in this way. Ten kilometres is a long way if there is no bus and you do not have a car. In geography we often measure distance in units of cost and time. Of course, some places are a long way however you measure the distance, for example northern Scotland from London.

Choice

There are many choices in transport: for example, should we go by car or train? It seems as if we have more choices than ever before. But should individuals be left to choose, or should governments make some choices for the whole of society? Many choices are likely to be controversial.

Public or private?

This is the largest of all the transport choices, and it is a choice for governments to make. Public transport, mostly buses and trains, is efficient. For example, more than 50 people can sit on a bus which occupies no more space than about three cars. But you have to wait for buses, sometimes in the rain, and they do not take you home to your door. So people have to be *encouraged* to use public transport, for example by charging very low fares. This costs the government money, and that has to be raised by taxes. In the end, it is the voters who make the choice!

A New Europe?

5·1   *Damned foreigners? This is a view that is sometimes held against European countries by British people. Is this attitude changing as we move towards a new Europe?*

How European are you?

Like the man in the picture on page 83, many people seem to have a negative attitude towards Europe. Others are keen to become citizens of the new Europe. Which group do you and your friends belong to?

The following questionnaire is designed to find out how European in outlook you are. The results might surprise you.

5·2 *The EC headquarters in Brussels.*

Questionnaire on Europe

1 Which of the following countriés are members of the European Community?
France   Ireland   Sweden   Greece
Egypt   Turkey   Spain   Austria   Poland

2a In which other country would you like to live? Choose one from the following:
Brazil   New Zealand   USA   France
Germany   India   Italy   Greece
Switzerland   Japan
b Give reasons for your answer.

3 Name the cities represented by the letters A–K on the map of Europe.

4 Are you *in favour of* or *against* the following?
a One type of currency (money) in Europe (the ECU).
b A European army rather than a British army.
c A European passport rather than a British passport.
d Britain becoming a state within a 'United States of Europe'.

5 Answer *yes* or *no* to the following questions.
a Do you speak a second European language?
b Would you be willing to learn another European language?
c Would you like to live and work in continental Europe?
d As well as being British, do you consider yourself to be European?

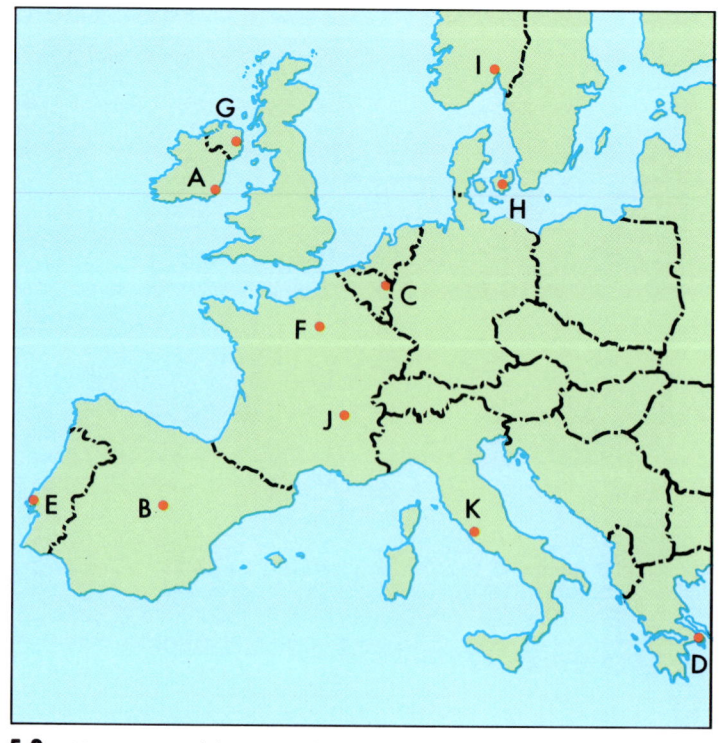

5·3 *Europe, and its capital cities.*

1 Each person in the class should first answer the questionnaire on their own.

2 Then for each question, collect the answers of the whole class. Your teacher will help you do this.

3 Display the information from the class results in graph form.

4 Describe how European you and your class seem to be from the results. Is this a good or a bad thing? Did any of the results surprise you?

Setting the scene

In 1952 several European countries joined together to regulate their coal and steel industries. Later, trade between these countries was encouraged, 'to bring an ever closer union among the peoples of Europe'. This grouping, of Belgium, France, Italy, Luxembourg, the Netherlands and West Germany, was called the European Economic Community (EEC). Since then the Community has grown to include many more countries, including the United Kingdom, and the group is now called the European Community (EC). It is likely to continue growing, and some East European countries could become members before the year 2000.

Originally all the countries that joined the EC agreed on the aims that were contained within the Treaty of Rome.

THE TREATY OF ROME

- More trade in goods and services between member states, without barriers.

- Free movement of people and money between countries.

- Help for farmers by paying higher prices and increasing the amount of food produced.

- Improvement of road and rail transport between countries.

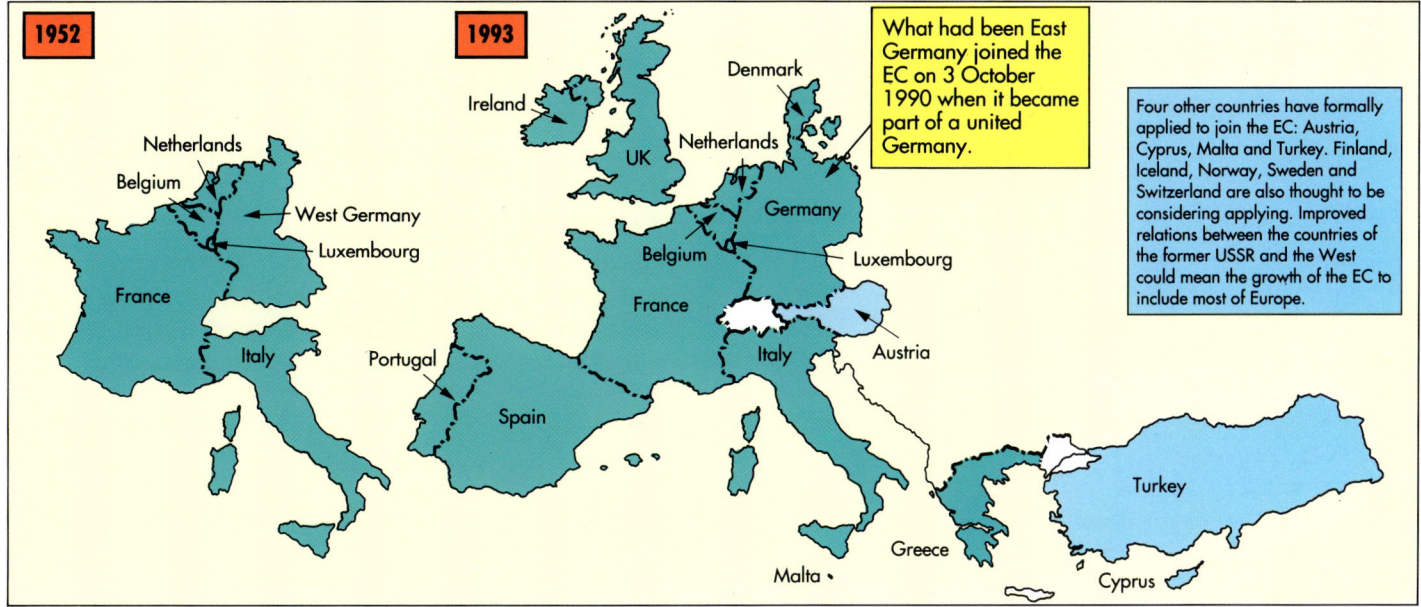

5·4 *The European Community: how the organisation has grown.*

European countries are gradually becoming more closely involved with each other. Step by step, progress has been made with farming and trade. Perhaps you have noticed the many different types of goods, including food, that we buy from within the European Community.

In 1986 the EC members agreed to the signing of the *Single European Act*. This stated that from 31 December 1992 there would be a 'single European market', and national borders within the EC would no longer matter for travel or trade. The British prime minister signed the Single European Act in 1986, but parts of the agreement have not been fully implemented. For example, the British want to keep passport controls, but Britain is alone in this wish.

However, for the 340 million people who live in the Community, this Act should allow any person to do the following anywhere within the countries of the Community.

★ Move freely without passport checks.	
★ Travel faster and more cheaply.	
★ Live and work anywhere in the European Community.	
★ Buy more and cheaper goods and services.	
★ Start building an office or factory.	
★ Produce goods and services in other European countries.	
★ Invest money in other EC countries.	

The Maastricht Treaty (1992)

In 1992 the leaders of all the EC countries met at Maastricht, in the Netherlands, to negotiate a new treaty. This treaty was to pave the way towards 'ever closer union'. Some leaders see the future emergence of a 'United States of Europe' – a *federal union of states* resembling the United States of America. Other leaders do not want this.

Later in 1992 the people of Denmark voted against the treaty in a *referendum*. The French also had a referendum which showed that the people were divided almost in half over the treaty.

In Britain the treaty has also caused fierce argument.

Ever closer *economic* union of Europe seems to attract support from almost everyone. But Maastricht, and the thought of closer *political* union, has caused many people to react. They want to defend their national interests and identity, which they feel could be submerged by a huge Euro-bureaucracy in Brussels.

The changes could be spectacular and have far-reaching effects on European life. The single European market will help the people of Europe to work more closely together, building upon the increasing friendship that has developed as a result of exchange visits between towns in Europe. By working together it is hoped that the people of Europe will become more wealthy, and the differences between the rich and poor areas of the EC will be evened out.

This unit is about how these changes in the EC might happen. We look at an area in France that is preparing for the single European market by facing up to the problems of the past and looking towards the new Europe.

The key questions in this unit are:

▷ Lille: a *European* city?

▷ How does France shape up?

▷ Towards Europe 2000?

▷ Lille: a *European* city?

In 1989 the former French prime minister, Pierre Mauray, said that 'the city of Lille is preparing for a great rendezvous with destiny'. Lille, the capital of its region Nord-Pas-de-Calais, has a particularly good geographical location within Europe. With a population of over 1 million, which includes the nearby towns of Roubaix and Tourcoing, this large urban area is set to become an important city in the new Europe.

  Lille is the biggest centre of employment in northern France. Most people work in banking, government, other service industries and retailing. The most important manufacturing industries are textiles, engineering, printing and food production (brewing). The centre of the city has many shops and offices but these have not spoiled the historical character of the area, which has many fine old Flemish buildings. There are many old, narrow streets in the centre, from which cars have been banned, and where there are flower markets, museums, shopping arcades and hotels.

5·5 *Lille town centre.*

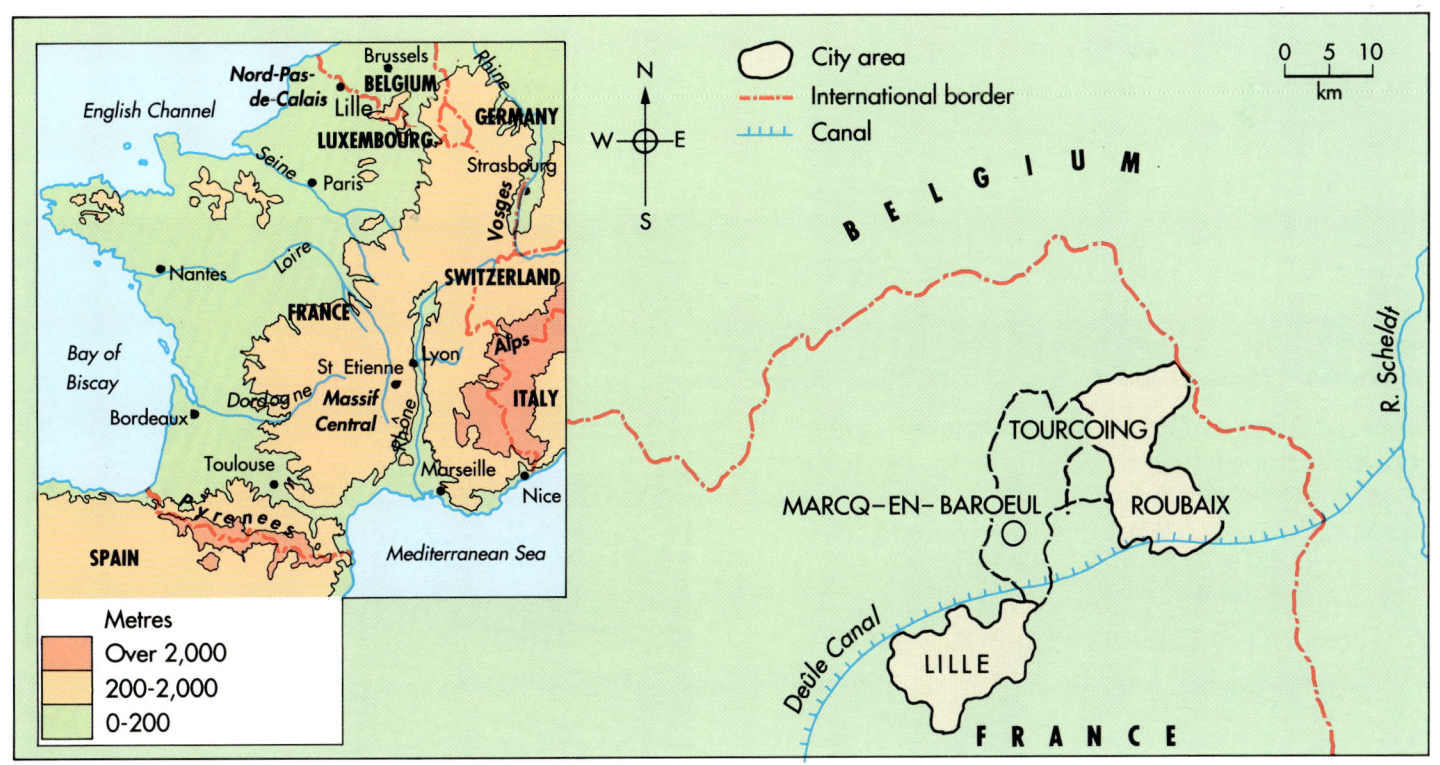

5·6 *France, and the Lille-Roubaix-Tourcoing urban area.*

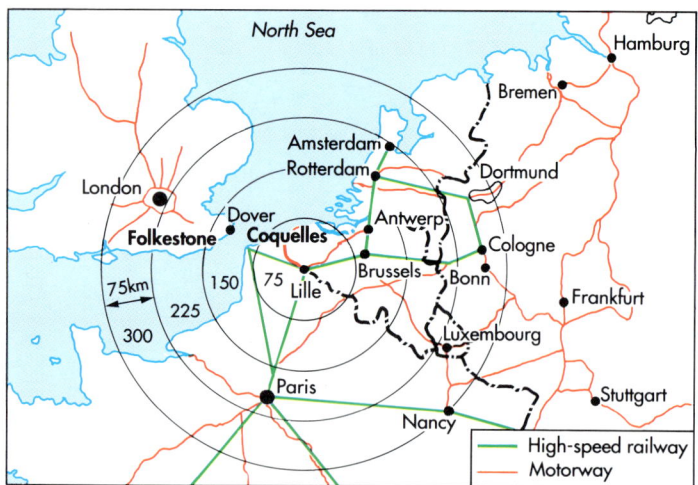

5·7 *The position of Lille in Europe, and the main transport routes across Europe.*

1   Lille lies close to the border between which two countries?

2   Name the six European capital cities that are within 300 kilometres of Lille.

3   Which important cities are connected by high-speed rail links in
a France
b Germany
c UK
d the Netherlands?

4   Estimate the distance between Lille and the centres of the following cities:
a London
b Paris
c Brussels
d Rotterdam
e Hamburg.

5   Suggest possible reasons why good links and high-speed travel are important to
a tourists
b business people
c companies that sell goods in other European countries.

6   Write a 150-word account to complete the following: 'Lille's location could be an advantage in a new Europe because . . .'.

The origins of Lille go back to 1066, when it was established as a trade centre between the countries of northern and southern Europe. The city is an inland port with docks on the River Deûle. The name Lille originates from the Latin word 'insula', meaning 'island'. The trade in river cargoes still goes on, and the city continues to grow and change. If you look at the map above, which has Lille at its centre, you can see that its location could be an advantage in a new Europe.

What is it like to live in Lille?

Josette Bossini is a 35-year-old economist who works for the new Lille underground system. She was born in Lille but has moved to Marcq-en-Baroeul, a suburb of the city. Her cottage was built more than 250 years ago and is situated close to the church of St Vincent's in the town square.

5·8 *Josette Bossini outside her home in the medieval street of Rue Raymond Derain.*

A short bus ride from Lille's railway station will bring you here to Marcq-en-Baroeul. With a population of 25,000, Marcq is big enough to be a town by itself. The people who live here work mainly in Lille, Roubaix and Tourcoing. These places are easily reached from Marcq because of the main roads and the motorways, including the A1 from Paris. The bus and tramway services are cheap and frequent.

Marcq-en-Baroeul gets its name from the River Marcq, and dates back a thousand years. Some parts of the town are old, even though most of the housing is modern. Today the main industry is still textiles, which employs 5,000 people. Other industries include metal-working, food production and farming; but most people work elsewhere in Lille. The main reason for living in Marcq is that it has many open spaces and good-quality housing, including some of the most expensive in Lille.

5·9 *Modern residential area of Marcq near the Collège des Hautes Loges.*

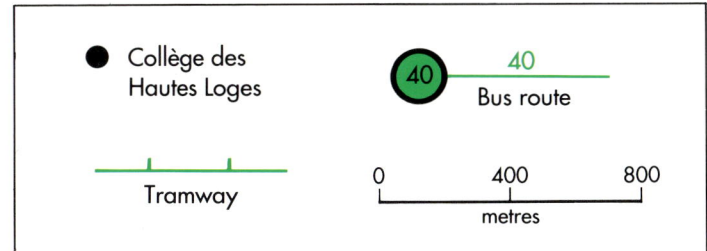

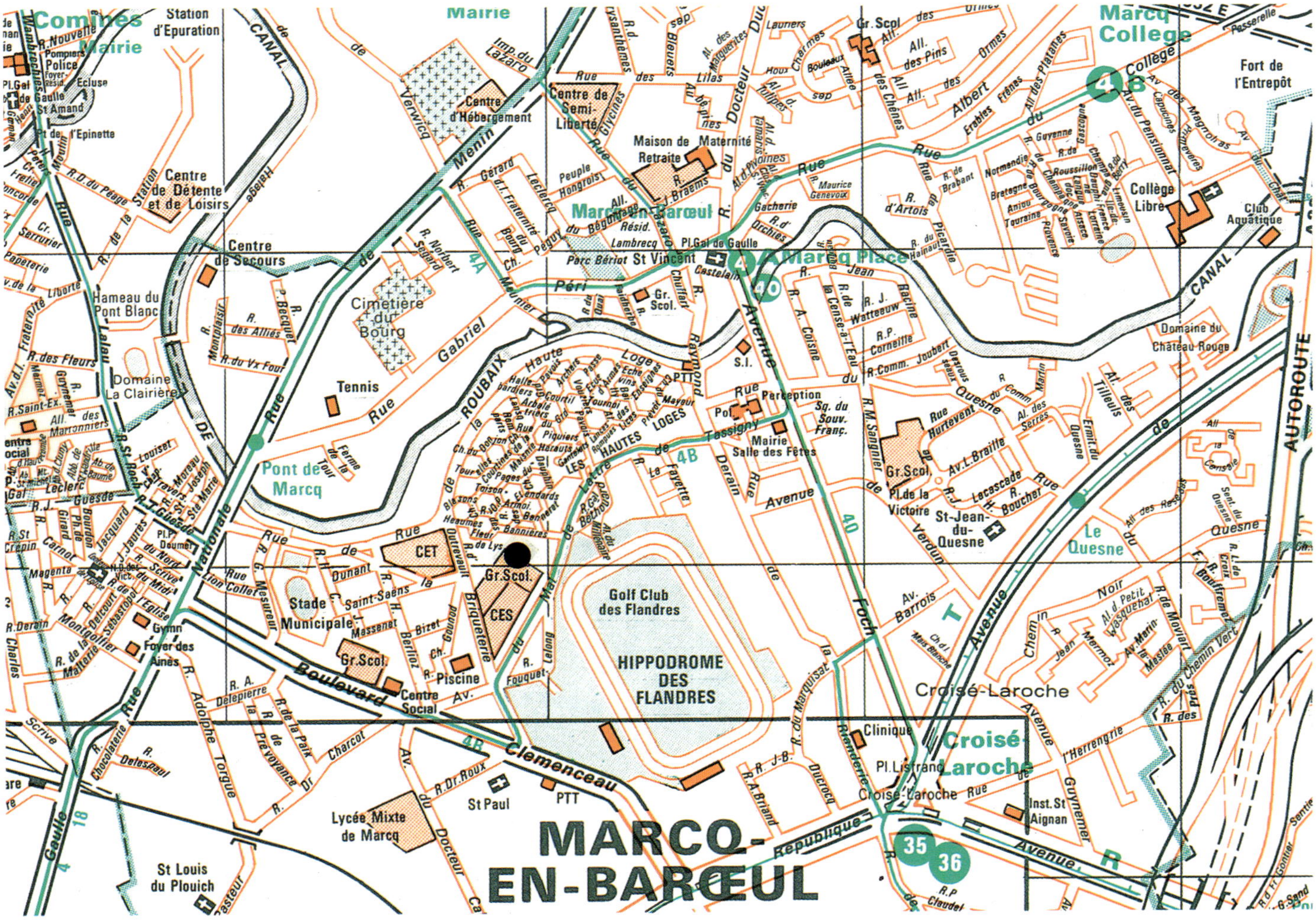

5·10 *The area of Marcq-en-Baroeul to the north of Lille city centre.*

Look at the map on page 89.

1 Using the key, work out the total *area* shown on the map extract.

Your calculation multiplies *x* and *y*; your answer should be in square kilometres.

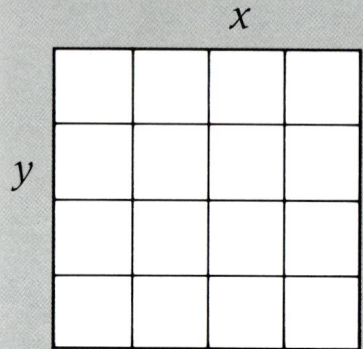

2 A piece of tracing paper will be needed for the following activity. Try to draw a *land use map* of the area shown in the extract.

A land use map is a generalised map; this means that a lot of the detail, such as individual roads or buildings, is lost. What is gained is a clear pattern of overall land use.

a On your tracing paper show the main *transport links*. This gives your land use map a 'skeleton'. Find and trace

- a motorway (autoroute)
- a canal
- main roads with bus routes
- tramways.

Use a different colour for each.

b Now mark on your tracing paper

- areas that you think are houses
- open spaces
- areas which contain services for the people, for example health, recreation, education.

Use a new colour or symbol for each.

Marcq in France, and Ealing in London

The people of Marcq are very interested in what is happening in other countries. So the town has 'twinned' with the borough of Ealing in west London. This will allow the citizens of both places to have greater contact with each other. Exchange visits between Marcq and Ealing will help people to find out more about their European neighbours and the countries they live in.

The Deputy Mayor of Ealing, and the Mayor of Marcq, meet in Ealing to discuss exchange visits between the people of their communities.

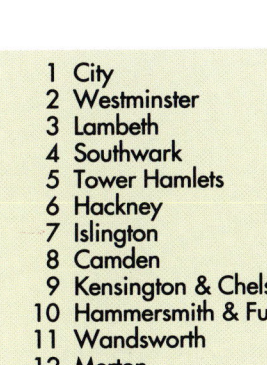

1 City	18 Bexley
2 Westminster	19 Newham
3 Lambeth	20 Barking & Dagenham
4 Southwark	21 Havering
5 Tower Hamlets	22 Redbridge
6 Hackney	23 Waltham Forest
7 Islington	24 Haringey
8 Camden	25 Enfield
9 Kensington & Chelsea	26 Barnet
10 Hammersmith & Fulham	27 Brent
11 Wandsworth	28 Harrow
12 Merton	29 Ealing
13 Sutton	
14 Croydon	30 Hillingdon
15 Lewisham	31 Hounslow
16 Bromley	32 Richmond upon Thames
17 Greenwich	33 Kingston upon Thames

5·11   *The London boroughs.*

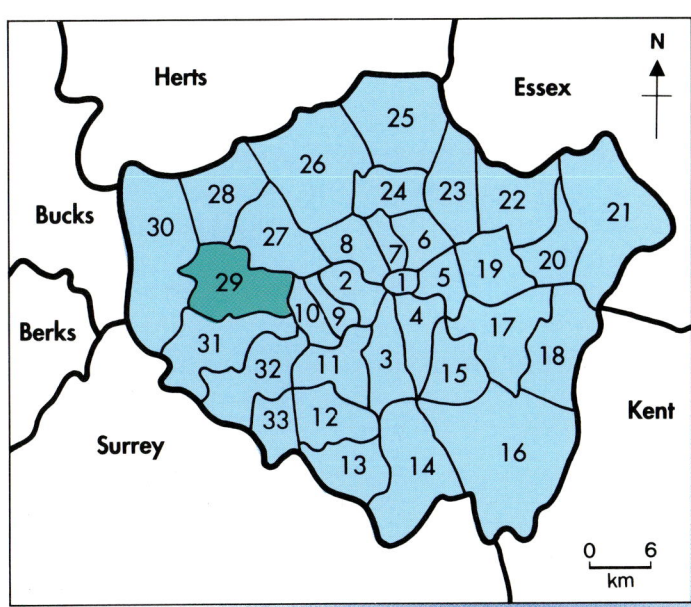

London

The original heart of London, the City has an area of 2.6 km². It has a resident population of 5,300, but more than 400,000 people work there.

In 1888 London was made a county – the county of London.

In 1965 it was expanded to become the area shown on the map, by absorbing the old county of Middlesex and parts of Surrey, Essex, Kent and Hertfordshire. This large area was administered by the Greater London Council (GLC).

In 1986 the GLC was abolished. London is now divided into the City plus 32 different boroughs.

It is a wonderful place, this London; a nation, not a city; with a population greater than some kingdoms and districts as different as if they were under different governments and spoke different languages.

Benjamin Disraeli, 'Lothair', 1870

1a Copy the following statements, and indicate whether they are *true* or *false*.
- There are not many leisure facilities in Marcq.
- Marcq has very good transport links with Lille and other cities nearby.
- Local industry is very important in Marcq.
- Marcq is a poor part of the city with old housing and much of it needing repairs.
- The River Marcq is very important to the town today.

b For each statement, give at least one reason to justify your answer.

2 Look again at the map on page 89. If you took a walk between the Collège des Hautes Loges and the church of St Vincent, you would see lots of streets with houses and flats, and some industrial buildings. Can you suggest *reasons* for these?

3a Produce a leaflet telling the people of Ealing about its twin town Marcq-en-Baroeul. Use written information, drawings and maps to give details of
- Marcq-en-Baroeul's location in France, and
- its main housing, employment, transport and recreation features.

b Suggest different twinning activities that could be arranged between the people of Marcq and Ealing.

c Find out if your home town is twinned with anywhere else in Europe. If it is, what kinds of twinning activity take place?

4 Ealing is part of London. Read the statement made by Disraeli about London over a century ago. Do you think the opinion he was expressing is more, or less, true today? Give a reason for your answer

▷ How does France shape up?

How much do you know about the geography of France? Did you know that it is the largest country in Europe (not counting Russia), and that it is twice the size of Britain?

France also has an important position in Europe. It shares a border with several other large European nations. It is an advanced, prosperous country which has much to trade with the rest of Europe.

France	
Population:	55.75 million
Capital city:	Paris
Area:	344,000 km²
Urban population:	73% of the total
Literacy rate:	95% of the adult population
Life expectancy (years):	
Males	72
Females	80
Families with	
Car	74%
TV	78%

France – a country of regions

How much do French children know about France's regional differences? At the Collège des Hautes Loges, a secondary school in Marcq-en-Baroeul, several 13-year-old pupils have drawn a map of France from memory and marked on it the main regions of the country. The pupils have also written short descriptions of some of these regions.

5·12 *The 22 regions of France.*

Geographical description of some regions of France

REGION A
A coastal area with a mild, wet climate suitable for dairying and market gardening. It contains the large port of Nantes, and 10% of the workforce are employed in fishing and farming. The fine coastal scenery and weather attract many tourists.

REGION B
A region whose capital is Strasbourg, to the east of the Vosges Mountains. The River Rhine marks the boundary of the region, and its industries are linked to wine, tobacco, textiles and engineering. The region has a continental climate.

REGION C
A southern region with a hot summer, often with a drought, and wetter, mild winters. Farming has improved recently with help from the EC. Industry has also become important in the Marseilles area. A popular tourist destination with resorts on the Mediterranean coast.

REGION D
A large upland area of volcanic rock. A variable climate, and mainly sheep and cattle are raised, as farming is difficult in many places. Some industry and, in the past, coalmining, may be found in St Etienne.

REGION E
The region contains the capital of France and is an important agricultural region, drained by the River Seine. It is a lowland region with some hilly outcrops where there is dairying, wheat and market gardening. There are some major industrial and service towns.

REGION F
A very flat region with arable farms and industrial towns, where textiles, engineering and coalmining were once important. This is France's most northerly region.

REGION G
A coastal region that is important for dairying and market gardening. It has many beaches and forests. There are good cross-Channel links with Britain.

REGION H
A region with some arable farming, but industry is the main economic activity. Heavy industry is in decline (iron and steel), but there is some light industry. Some people work in Germany, which is close by.

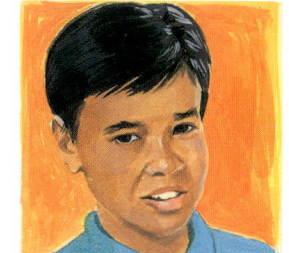

NORD - PAS - DE - CALAIS
The Nord-Pas-de-Calais is a beautiful region.
It is very flat and green and borders on the
North Sea. I like it a lot - I live here!
Nicholas Derain

PROVENCE
Large numbers of people visit the region,
especially in the summer. The Mediterranean
Sea is very polluted and the weather is hot.
Some people are unfriendly, but the landscape
is beautiful. There are many yachts which
sail in the sea.
Caroline Blanc

PARIS BASIN (centre)
The River Seine flows through Paris with its 8.5 million people.
The air in the city is very polluted because of the large
numbers of cars. The weather is very grey, but it can
be a beautiful place.
Karima Zahnjan

LORRAINE
Lorraine is a region in the east of France. The people
there have their own accent. There is quite a lot of
unemployment.
Daniel Laurant

Barbara Pyke

ALSACE
Alsace is a mountainous region but quite
flat where it has a border with Germany.
It has good weather. There is much sunshine
in summer and snow in winter, when
it is possible to ski.
Yves-Gabriel Thollon

BRITTANY
Brittany is a magnificent region bordering on the Atlantic.
It is reached by crossing the River Seine. It is a region where
flowers and plants flourish. It has 2 million inhabitants.
Some work as fishermen in the sea ports.
Audrey Doom

Daniel Laurant

MASSIF CENTRAL
LANDSCAPE: Mountainous
SPECIAL FEATURES: Extinct volcanoes, known as
puy de dômes
POPULATION: 5,000
Jonathan Cabusat

NORMANDY
Ah, Normandy! What a beautiful region
of our country. Its beaches, its cows,
its apples! In spite of the region's
tendency to be wet, it is a beautiful
region.
Olivier Decoste

5·13 *Descriptions of some of the regions of France by pupils of
the Collège des Hautes Loges.*

1 Examine the French pupils' maps of France on page 93.
a Which map shows the most accurate shape of France?
b Name the regions that have been (i) correctly and (ii) incorrectly drawn.
c Which map shows the regions of France most accurately?
d What do you notice about the accuracy of Barbara Pyke's regions as distance increases from Nord-Pas-de-Calais? Suggest a reason for this.
e What is unusual about the size of Nord-Pas-de-Calais on Barbara's map? Why might she have drawn it this way?

2 Using the pupil descriptions on page 93, and the map of France and description of some regions on page 92, name the regions A–H.

3 Which pupil gave (a) the most accurate and (b) the least accurate description of a French region? Give at least one reason for your answers.

4 Which pupil gave the most *interesting* description?

5a On your own map of France, mark on each region A–H.
b Create a 'regions file', giving a summary for each region, using the descriptions on page 92.

6 Make a list of what you think might be
a the three richest regions, and
b the three poorest regions of France.

France's regional problem within Europe

France has a varied geography. This means that some regions of the country have an advantage over others. The richer parts of France have a large number of modern industries, new office and other service jobs. These areas employ many people and provide many goods and services, which bring much wealth to the people who live there. Paris (Centre) is an example of this type of region.

Some areas of France are poor because they have old, outdated industries. Other regions have little industry and people rely on farming for work. If these areas are far away from richer areas, this can add to their problems. One of the newspaper headlines here highlights a problem often found in poor regions – unemployment.

The most wealthy areas of Europe are in an area called 'the Golden Triangle', although not all areas within the Golden Triangle are rich, and the Nord-Pas-de-Calais has a high level of unemployment.

Chômage : de nouveaux écarts dans le Nord – Pas-de-Calais

NOTRE RÉGION

No. 55 Mars 1991 10 F

Le mensuel d'information du Nord et du Pas-de-Calais

La réforme qui trouble le jeu électoral pour les Régionales

(Pages 14 et 15)

Le coût de la fortune

Notre région est l'une de celles où le montant de l'impôt sur la fortune rapporté au nombre d'habitants est le plus élevé: près de 52 francs en 1990. Seules l'Ile de France (de loin), Provence-Alpes-Côte-d'Azur et Rhône-Alpes (de plus près) font mieux.
Le département du Nord, avec 70 francs par habitant en 1990 recèle davantage de richesse que le Pas-de-Calais. Il vient même au neuvième rang de tous les départements français et au cinquième si l'on exclut l'Ile de France.

(Page 6)

ENQUÊTE D'OPINION:

Ce qu'on pense dans la région

5·14 *From 'Notre Région', a local newspaper of the Nord-Pas-de-Calais. What does it say about unemployment, and about 'the cost of wealth'? You can find out by turning to page 99.*

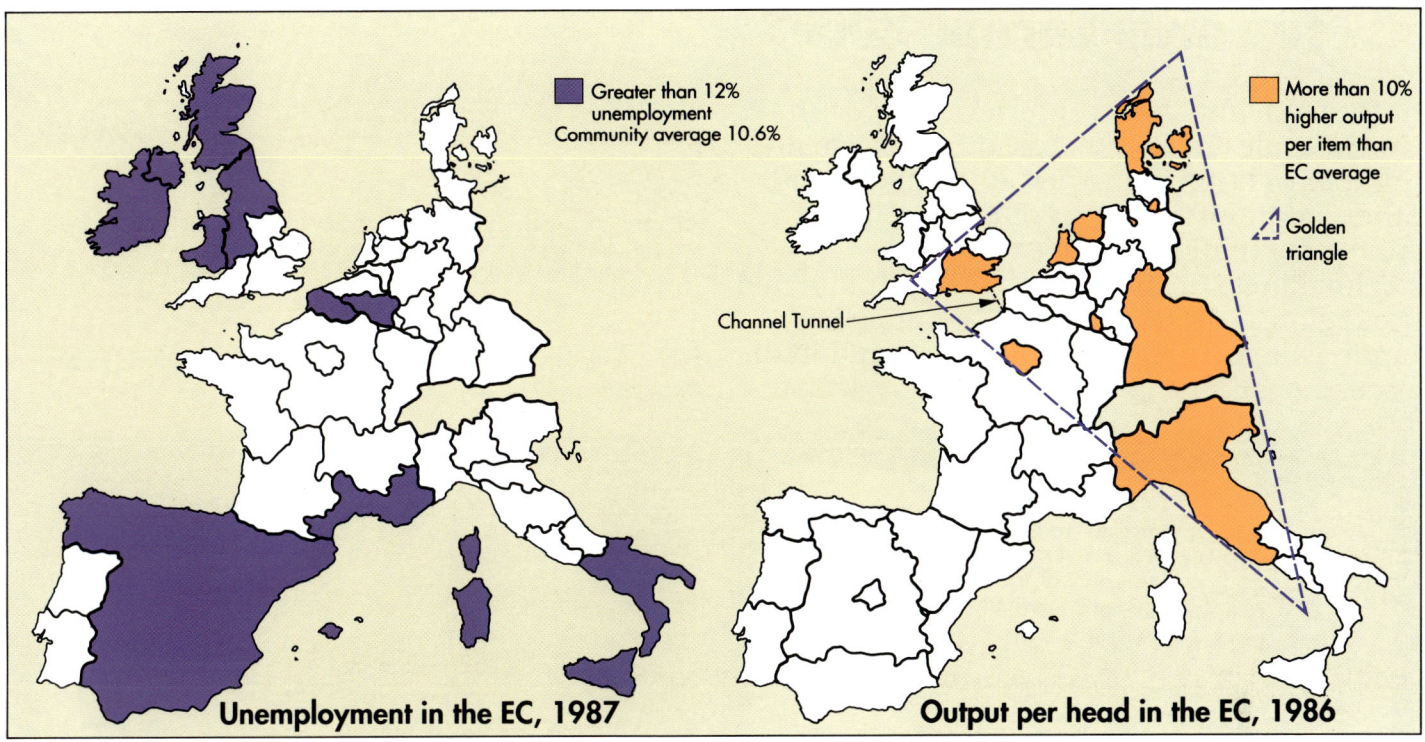

5·15 *Rich and poor areas of the EC, using two different measures.*

The EC gives aid to regions if they
- suffer from a slow rate of economic growth
- have many declining industries
- have high unemployment
- use poor farming practices.

This is done through the European Regional Development Fund (ERDF), which was set up to solve these problems. All members belonging to the EC give money to the fund. In 1987, France received 9% of the fund, and Nord-Pas-de-Calais was given financial help to attract new businesses. The money was used for the following:
- Provision of land and buildings for new businesses.
- Training projects to help new businesses, especially small ones, to start up.
- Improvement of the environment by clearing derelict land.
- Encouragement of tourism based on the historical and cultural attractions of the area.
- Improvement of the transport system.

The Nord region also suffers from being a border area, as in the past this has complicated travel and trade between France and Belgium. The EC now gives grants (money) for training schemes in both countries, so that Belgian and French people can work more easily in each other's country. The Single European Act makes this even easier.

---

1 What do you think the phrase 'France has a varied geography' means?

2 Name the regions of France that have more than 12% unemployment.

3 Name the richest region of France.

4a What is 'the Golden Triangle'?
b Which do you think are the wealthiest parts of this area? (An atlas may help you to answer this question.)

5 Look at the area of the Golden Triangle in the map above. Find the region of Nord-Pas-de-Calais. Why do you think the Mayor of Lille said recently, 'Lille now finds itself in the right place at the right time'?

▶ Towards Europe 2000?

In the 19th and 20th centuries, the wealth of Nord-Pas-de-Calais was based upon a small number of large industries. In common with other old industrial areas of Europe, these were shipbuilding, coal, textiles and steelmaking. During this time they dominated the region. A famous French writer called Emile Zola made this clear in a description of a scene near Lille at night during the 1870s.

	1954	1962	1968	1985
Coal	143,600	112,100	87,400	8,500
Steel	30,200	37,600	35,600	33,000
Textiles	170,000	140,000	121,000	95,500
Cars	1,900	5,800	4,500	37,500

5·17 *Employment change for a selection of industries in the Nord region, 1954–85.*

> Etienne now commanded a view of the whole district . . . and all he could see was distant furnaces and coke ovens with a hundred chimneys making sloping lines of crimson flames; whilst to the left the two blast furnaces were burning blue in the sky like monstrous torches. It was as depressing to watch as a building on fire. As far as the threatening horizon the only stars which rose were the fires of the land of coal and iron.

From E. Zola, 'Germinal', 1885

Evidence of Lille's industrial past can still be found today. The mills of Motte-Bossut produced textiles in Tourcoing until 1981, but today lie idle. Much of the old industry of the Nord has disappeared and also the jobs that supported the people, their families and the whole community. The decline of these industries made the Nord-Pas-de-Calais an

5·18 *An old textile factory in Marcq-en-Baroeul.*

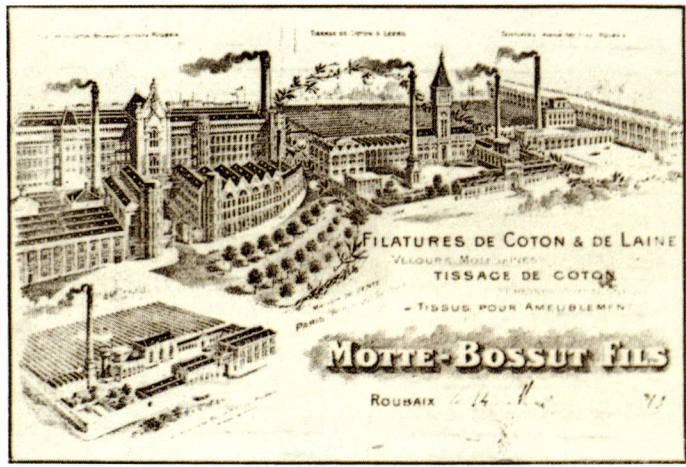

5·16 *The Motte-Bossut mills were centres of heavy industry until cheaper textiles produced overseas forced them to close.*

unattractive area for new businesses. This is beginning to change as Lille now finds itself 'in the right place at the right time'.

For 15 years Lille has been planning to improve its attractiveness for businesses. Since 1983 the city has been using a fully automatic underground system with driverless trains, called the VAL. Old buildings have been restored, and more tourists now visit Lille. There are many facilities in the city such as opera, theatres, museums, sports centres and golf courses, which make Lille an attractive place to live in. By 1991, F450 million (£45 million) had been spent on the main roads and motorways to allow better communications within the Lille area, and to other parts of France and Europe.

1 What were the main industries of the Nord in the 19th century?

2 Describe the scene in the illustration on page 96 (bottom).

3 Suggest what the link is between the textiles industry and the coalmining in this area.

4 On one graph, draw four lines to show how employment has changed in **a** coal, **b** steel, **c** textiles, and **d** car manufacturing between 1954 and 1985.

5a What two changes affect Lille's position in Europe after 1992?
b Describe the problems that Lille has suffered from in the past.
c How will the Channel Tunnel affect Lille's position?

Communications and transport

Improvements in transport and communication are a major part of Lille's attempt to attract new businesses to the region. As well as changes to the road and underground systems, a major railway station is being built in the city centre for the new TGV-Nord railway line. This new line will link up with the other major cities of Europe, and Lille will act as an interchange between France, Britain and the rest of Europe (see the map on page 88). The high-speed TGV (*Trains à Grande Vitesse*) reach a maximum speed of 300 kilometres per hour, and will considerably reduce travelling times between European cities.

5·19 *Major road improvements, such as a new traffic system near La Madeleine, provide easier access around Lille.*

	1991			1993	
Victoria (London)	dep.	12.00	Waterloo (London)	dep.	12.00
Dover	dep.	14.00	Folkestone	dep.	12.30
Calais	arr.	15.40	Coquelles	arr.	13.05
Lille	arr.	18.15	Lille	arr.	14.00
Total		6 hr 15 min.			2 hr

5·20 *Train times between London and Lille before and after the building of the Channel Tunnel and the TGV-Nord line.*

The new station is being built on unused land to the north of the present railway station. The entire project will be built in the heart of the city without the need to demolish any buildings. The 150-hectare site will include a business centre with the most up-to-date communications systems.

The cost of the project will be F800 million (£80 million), which is being paid for by the French government, the city council and the EC. This is an example of how money from the ERDF is spent to help the poorer areas of Europe. The development should make Lille a very attractive place for businesses, with a pleasant environment to live and work in.

5·21 *The TGV – the world's fastest train – should help to bring business to Lille.*

5·22 *The building of the TGV station at Lille. In the background is part of the tramway system.*

Work in small groups for this activity. First, read this:

In Emile Zola's day, Lille was typical of a heavy industrial town located on a coalfield. Conditions were similar to those found in towns on other coalfields in Europe in the 19th century. Gradually, over the last hundred years or so, conditions have changed. Now, at the end of the 20th century, the city of Lille knows that its future is not with coal or 'heavy' industries like steelmaking or engineering.

Person 1

You are a city planner in Lille. Your name is Roland Rickard. You have been invited to a meeting in *Katowice*, in Poland. Poland is one of several East European countries that would like to join the EC. Your fare has been paid by the European Development Fund (EDF) because the EC thinks your visit could be useful. Your experience of Lille, and knowledge of the changes there over recent years, may be something many towns in Poland could learn from.

• Your role is *enthusiastic persuader*.

Person 2

You are a city planner in *Katowice*, in Poland. Your name is Lech Mankowicz. You meet Roland Rickard at the EC meeting held in your home town. You are very keen to hear about Lille because Katowice has a similar but more recent history: there are many thousands of unemployed miners and factory workers – just as there were in the past in West European towns like Lille. You are not sure, though, that the Lille story can be copied by Katowice. For one thing, isn't Lille in a more favoured geographical location than Katowice?

• Your role is *sceptical critic, who is in need of help*.

1 After reading the first part of this activity, decide who is going to take each role. If there are more than two people in the group, the others will be observers and recorders.

2 Prepare your role. Think ahead to your meeting. What are you going to say at the meeting?

3 Have your meeting. You may read out statements to each other. Then you can have a conversation.

4 The observer(s)/recorder(s) should make notes on the meeting. You are not allowed to say anything until the end. Then you can make your report, as if you were writing a *press release* for the newspapers.

Unemployment: new differences in the North – Pas-de-Calais
(Pages 6 and 7)

OUR REGION

No. 55
March
1991
10F

The monthly information magazine for the North and the Pas-de-Calais

The reform that is disrupting the electoral game for the Regionals
(Pages 14 and 15)

The cost of wealth

The increase in tax on wealth per head in our region is among the highest in the country: almost 52F in 1990. Only l'Ile de France (further afield), Provence-Alpes-Côte-d'Azur and Rhône-Alpes (closer to hand) exceed this.

The North *Département* receives more wealth than the Pas-de-Calais, with 70F per head in 1990. It was even ranked ninth among all the *départements* in France, and fifth when l'Ile de France was excluded.
(Page 6)

OPINION POLL
What we think in the region

5·23 *From 'Our Region'. English version of the newspaper item on page 94.*

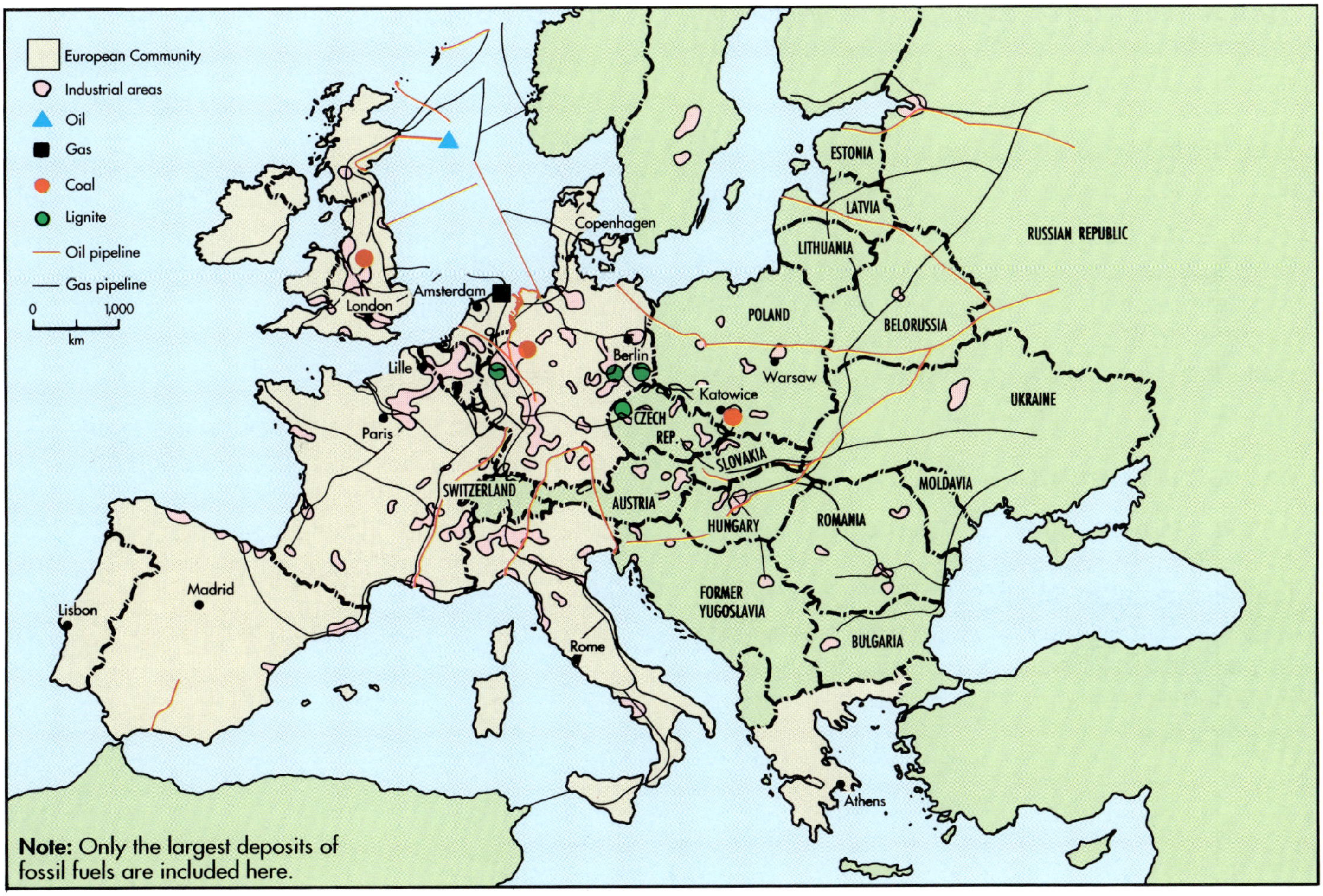

European Community
Industrial areas
Oil
Gas
Coal
Lignite
Oil pipeline
Gas pipeline

0   1,000
km

ESTONIA
LATVIA
LITHUANIA
RUSSIAN REPUBLIC
Copenhagen
POLAND
BELORUSSIA
London
Amsterdam
Berlin
Warsaw
UKRAINE
Lille
Katowice
Paris
CZECH REP.
SLOVAKIA
MOLDAVIA
SWITZERLAND
AUSTRIA
HUNGARY
ROMANIA
Madrid
FORMER YUGOSLAVIA
Lisbon
Rome
BULGARIA
Athens

Note: Only the largest deposits of fossil fuels are included here.

5·24 *A new Europe.*

European unity

This is the idea that the different countries and peoples of Europe should understand and appreciate one another. This might include encouraging a positive attitude towards different languages, cultures and lifestyles, in order that people might work more closely with and visit their European partners. The idea behind the creation of the EC was to help produce this kind of unity. At the same time it has provided a very rich and powerful base to compete with the USA and Japan.

Distinctiveness of place

The many features of a place make it unlike any other. For example, the suburb of Marcq-en-Baroeul has housing, industry and community buildings which are there because of its location on the River Marcq. Living there is not quite the same as living anywhere else in the world.

Industrial decline

This happens when an area loses its industry over a long period of time because it is no longer profitable. Many factories in one or more industries may close down because they are old and inefficient, or because they are suffering from competition from other regions or countries.

Communications revolution

This is about the improvements made in transport (road, rail and air) to link up more places and allow faster and more convenient travel between places. The ease of travelling between one city and another by train after the building of the Channel Tunnel is an example of this.

Regional policy

Regional policies are made by governments, acting alone or together. They are trying to solve problems in an area, such as unemployment, bad housing and low income, by improving its economy. One example is the EC providing money to train people to work in more skilled jobs so that industry will be attracted to an area.

Contrasting regions

This term describes how the parts that make up a country can be different from each other. They can be different in terms of area, population, climate, industry, and so on. A country as large as France has several contrasting regions.

A World Full of People

At every beat of your heart . . .
. . . two babies are born.

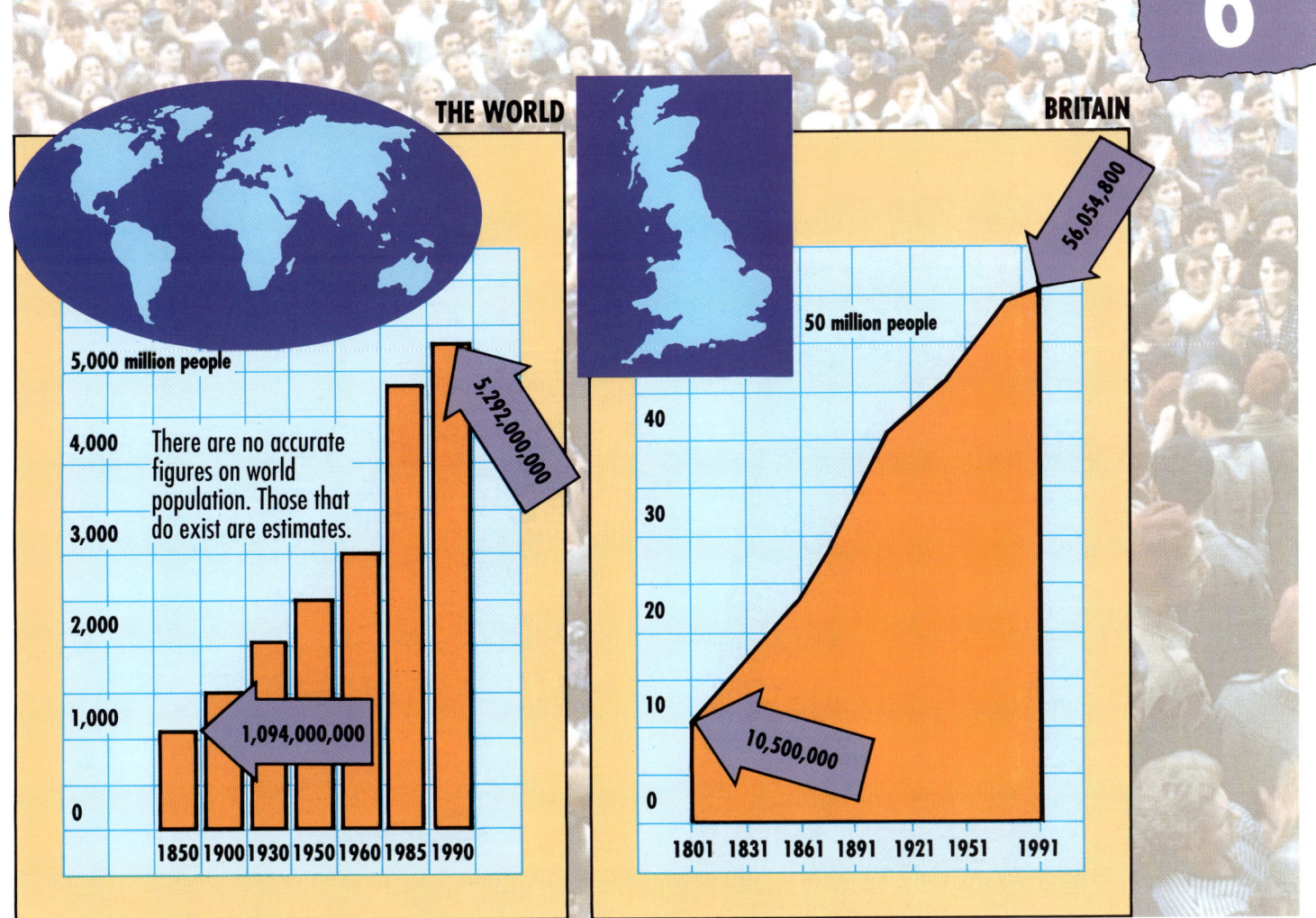

THE WORLD

5,000 million people

4,000 — There are no accurate figures on world population. Those that do exist are estimates.

3,000

2,000

1,000

0

1850 1900 1930 1950 1960 1985 1990

5,292,000,000

1,094,000,000

BRITAIN

50 million people

40

30

20

10

0

1801 1831 1861 1891 1921 1951 1991

56,054,800

10,500,000

6·1 *A growing population: during the last 10 years, the world's population has increased by* **750 million***. What will be the population of our planet by the 21st century?*

101

What does the word 'population' mean to you?

1 Write down a series of words and phrases that you associate with the word 'population'.

2 Compare your ideas with the person next to you. Do you have the same ideas? Discuss similarities and differences.

Your class is part of the school's population.

3 What is the population of your class?

4 Think of various ways of dividing up the population of your class into different categories, for example boys and girls.

Counting people

A *census* is carried out in Britain every 10 years, to count the number of people in each household on one particular day. The last census was on 21 April 1991.

6·2 *Information explaining the reasons for holding a census.*

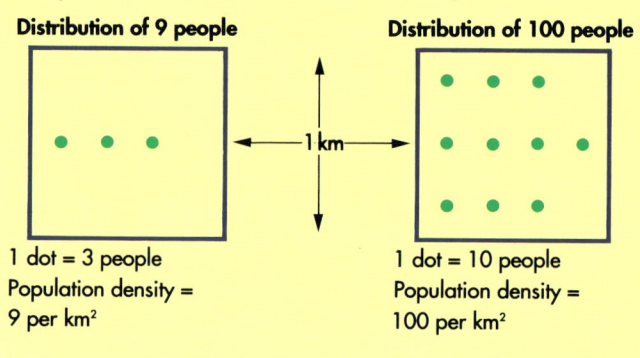

Population density

Population density is *how thickly spread* the population is. In these diagrams, each square represents an area 1 km x 1 km, or 1 km².

Distribution of 9 people

Distribution of 100 people

1 km

1 dot = 3 people
Population density =
9 per km²

1 dot = 10 people
Population density =
100 per km²

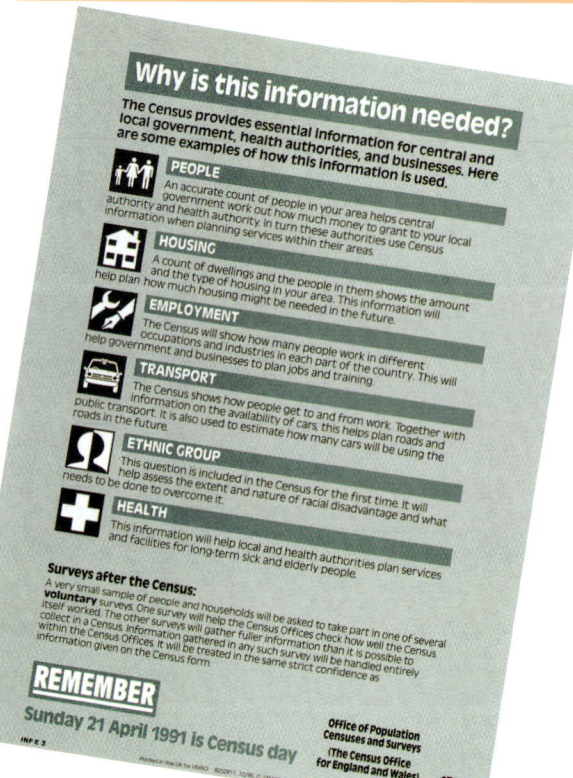

Doing a census survey

1　During your geography lesson, carry out a census to find out the distribution of the school's population. Each group will need a plan of the school, or a table listing the room numbers and other areas like the hall, on which to collect the information.

Place	No. of people
Room 1A	
Room 4D	

2　Once you have collected the information, think about ways of presenting it. One way is to show distribution using a dot map. Each dot represents a certain number of people. Show the school's population distribution on a plan of the school using the dot map method.

3　Discuss in your groups why it is useful to have categories of the population, such as girls/boys, or different age groups. (Look at the National Census information opposite.) Design an information sheet to tell students about a school census.

4　As a class, discuss whether you can make generalisations about
- population based on your class survey
- factors affecting population distribution based on your school survey.

Setting the scene

Planet Earth is 4,600 million years old

Imagine that Earth is a person who is now 46 years old. Nothing is known about this person's first seven years, and in fact we know very little about what happened in Earth's life until middle age, when Earth began to flower. Dinosaurs and the great reptiles were not known until Earth was 45. Mammals arrived only eight months ago, and in the middle of last week, human-like apes changed to ape-like humans. Real humans have only been around for four hours. During the last hour, humans discovered agriculture. The industrial revolution began a minute ago, and during those 60 seconds humans have made a rubbish-tip of Paradise. They have multiplied to plague proportions, causing the extinction of 500 species of other animals, they have ransacked the planet for fuels, and effectively destroyed this oasis of life in the solar system.

Adapted from a Greenpeace leaflet

This extract paints a gloomy picture of the devastating effect that humans have had on the world in a *very, very short period of time*.

6·3　*The story of Earth.*

103

1a Make a list of items you use in your everyday life.
b Try to group these into
- those that are absolutely essential *needs*, and
- those that are '*wants*' rather than needs.

c Compare your list with that of another person in the class. Discuss your decisions, and arrange them by drawing two overlapping circles.

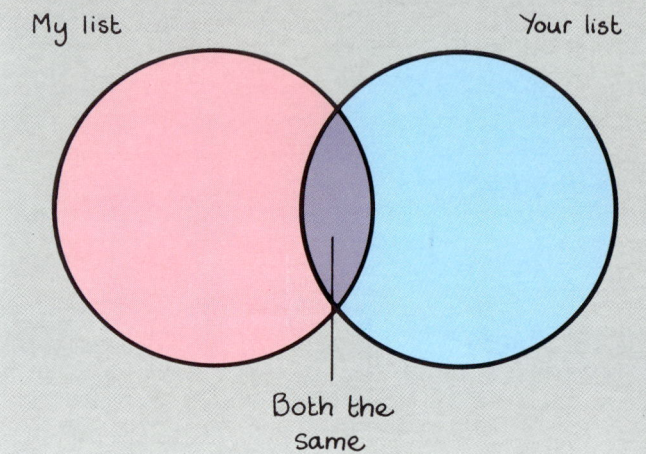

My list   Your list

Both the same

2a You live in the rich part of the world and most probably live in a town or city (an *urban area*). In many poorer parts of the world, children your age will live in the countryside (*rural areas*). Go through your lists and see whether you think these are the same wants and needs as those of a person your age in a *rural* part of India.
b What do you think should be your basic rights, whether you live in Britain or in India?

3 Design a poster with the title, 'Your rights – whoever you are, wherever you live', to show the issues you have discussed.

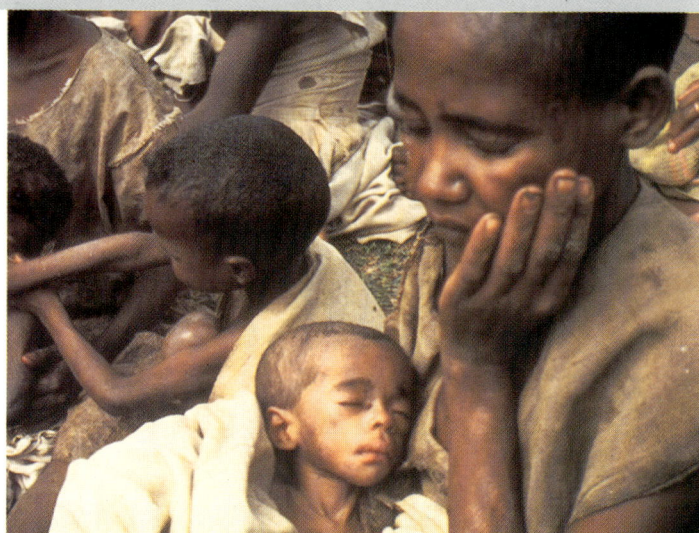

The birth of a baby in the United States imposes more than 100 times the 'stress' on the Earth's resources and the environment as a birth in Bangladesh. A baby in Bangladesh does not grow up to own a car or eat grain-fed cattle. The lifestyle in Bangladesh does not require huge amounts of minerals and energy. An average US family affects the environment 40 times more than an Indian family, and 100 times more than a Kenyan family. Humans are changing the world in which we live every day. (See Unit 8, 'Consuming the Earth's Resources', in *Green Pieces*.)

At current rates of destruction, much of our tropical rainforest will be gone within 25 years. Gone too will be at least one million species, extinct and lost for ever. More directly, some 50 million square kilometres and nearly a billion people are affected by *deforestation*. Because these people suffer from a shortage of firewood (their main fuel), or their crops are reduced as a result of soil erosion, many are said to be *living on the margins*. 'Margin' means edge – in this case, the edge of decent existence.

When famine is a constant threat

Famine facts

■ According to the World Health Organisation, 40,000 children under the age of 5 die from disease, malnutrition and neglect in developing countries every day. That is *15 million children a year*.

■ Worldwide, almost 750 million people suffer from hunger and malnutrition. That is *more than double the population of Europe.*

■ According to the United Nations Food and Agriculture Organisation, between 1973 and 1983 the area in which cereals and root crops were grown in developing countries increased by 8 per cent. The area for growing cash crops such as coffee, sugar and soya beans increased by 48 per cent.

6·4 *The facts of famine.*

6·5   *The burdens of Earth.*

The planet's heaviest burden

IT WAS courageous of the Prince of Wales to put population control at the heart of his long speech yesterday. We have to ask ourselves, he said, whether we can continue to ignore the prospect of a virtual doubling of the world's population, to somewhere approaching 10 billion, by 2050. He identified population growth in the Third World, in which 84 per cent of humanity would be living by 2020, as the problem underlying the world-wide degradation of the environment. "We will not slow the birth rate until we address poverty," he went on. "And we will not protect the environment until we address the issues of population growth and poverty in the same breath."

The diagnosis is correct: poverty is the greatest environmental menace. Yet Prince Charles was clearly in some agony of mind over how to overcome it without dangerously accelerating the depletion of natural resources.

From The Independent, 20 May 1992

Population is a topic that worries many people. It is a complicated matter and we need to understand it better. The items on these two pages show both the *worry* (is it too late?) and the *complications* (how can the world get rid of poverty?) involved. In this unit we consider three key questions:

▷ Where is the world's population?

▷ The population explosion – fact or fiction?

▷ The seesaw: How can the world balance population and resources?

▷ Where is the world's population?

Where do people in the world live (population *distribution*)? The population *density* (number of people per square kilometre) is much greater in some parts of the world than others. We look here at why this is so.

1 You will need a large copy of this small world map. Notice that the map is divided into many small squares. Look at the map and try to decide where most people live. Shade in up to 100 squares to show where you believe most of the world's people live – that is, areas that you believe have a *high population density*. Empty squares are the areas where you think few people live (*low population density*).

2 Try to give reasons for the decisions you made when shading in your world population map. Compare your distribution and explanations with those of your neighbour.

3 Using the map opposite of world population distribution, and an atlas map showing the countries of the world to help you, write a description of the world's population distribution. These questions may help you:

- Which countries are densely populated (have a high population density)?
- Which countries are sparsely populated (have a low population density)?
- Which continent is the least populated?
- Which continent is the most populated?
- Which hemisphere is the more densely populated: the northern hemisphere or the southern hemisphere?

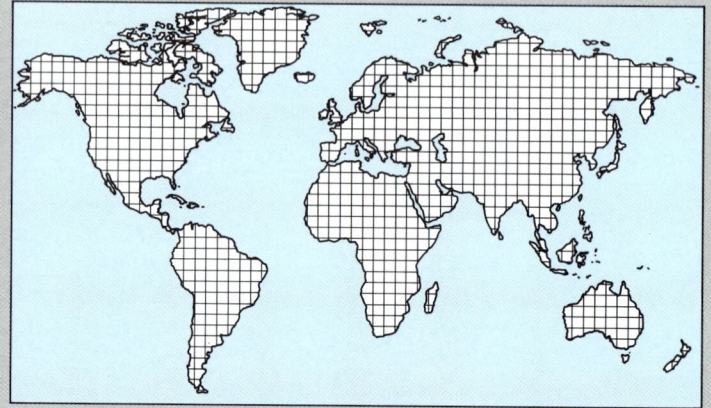

6·6 *Map for showing world population distribution.*

4 You can now compare *your* map with the map opposite.

a Using an atlas, identify and name areas of countries or continents where you shaded squares wrongly. In other words, areas which you guessed are densley populated but which really have a low population density.

b Identify and name areas of countries or continents where you chose squares correctly.

c Using a map which shows the world's major physical features, try to find reasonms for variations in population density. You could write two columns, one headed 'High densities' and the other 'Low densities'.

5 Now draw a proportional symbol map to display the figures presented in the table (left), for the continental distribution of population.

On a map of the world, draw a simple symbol (for example a human figure) in each continent. The height of the symbol should indicate the population of that continent. Use the scale 1 mm = 20 million people. This scale must be clearly displayed on the map. Give your map a title.

Continent	Population (millions)
Africa	498
Asia	2,671
North America	256
South America	378
Europe	488
Oceania	24

Note: *Antarctica is not included for this exercise.*

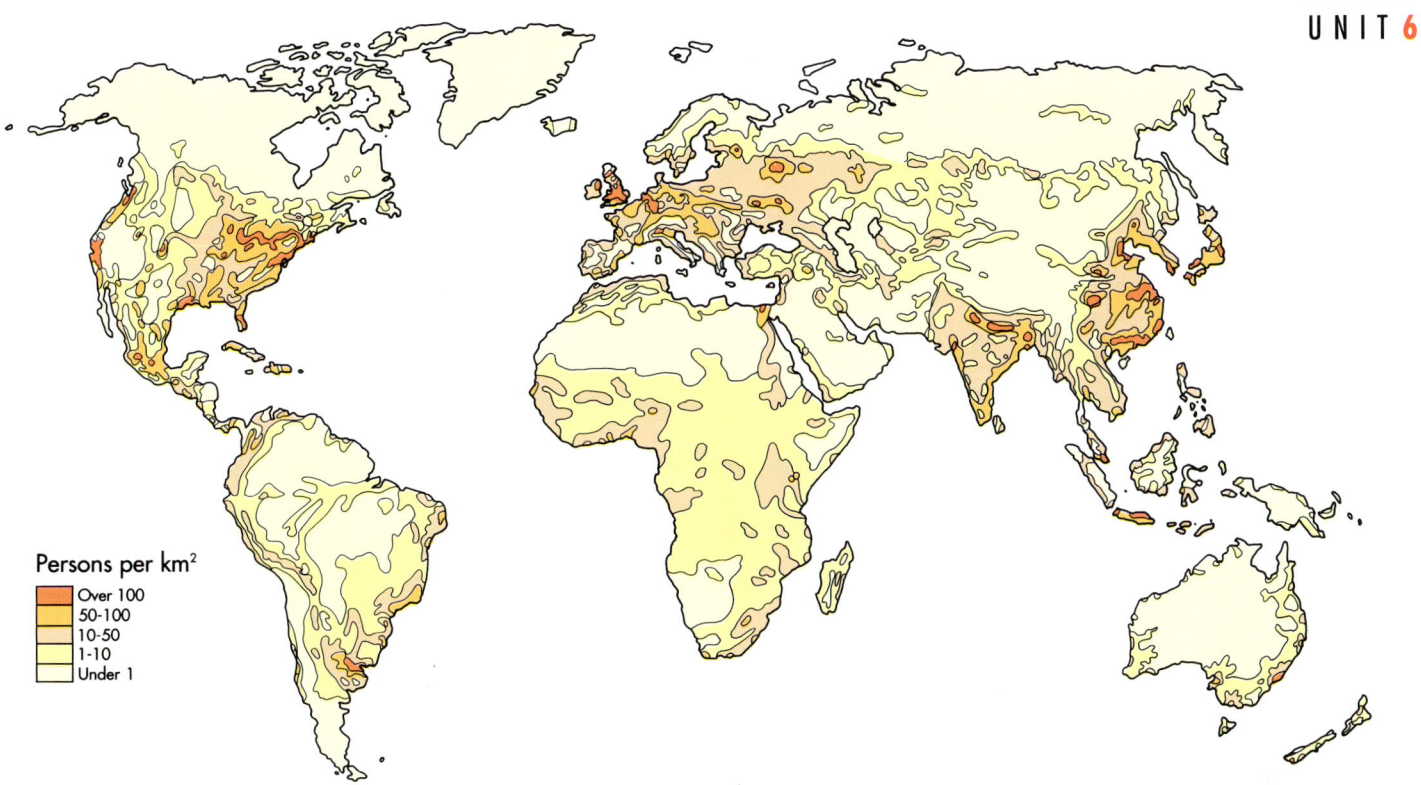

6·7 *World population distribution.*

Explaining the patterns on the map

Are people grouped together in certain parts of the world by chance, or are there reasons for the densely and sparsely populated regions of the world?

The reasons can be *physical*. For example, 93% of Egypt's population live within a very short distance of the River Nile, because it is almost the only source of fresh water and fertile soil in that country.

Reasons can also be *human*. For example, many great cities have grown up near the coast or at the mouth of a large river because trade and industry have been able to flourish there, providing jobs and supporting many people.

When we look for reasons for the world population distribution in physical geography, we are certain to find *anomalies*. This is because people are not governed only by the physical world. People can make unexpected decisions about where to live.

Anomalies

An anomaly is *something that does not follow the expected pattern.*

What's *normal*?

Things we expect – the usual things that happen, 'on average'. For example:

– Men are taller than women.

– Tobacco causes premature death.

What's *abnormal*?

Things we don't expect, the unusual, things that happen as exceptions. For example:

– A coachload of tourists in which the women are taller than the men.

– The 95-year-old who has smoked since the age of 15.

Exceptions such as these are *anomalies*.

107

▷ The population explosion – fact or fiction?

Today, people sometimes say that parts of the world, or some countries in particular, are experiencing a 'population explosion'. What does this mean? After a very long period of slow but steady growth, the population has recently risen dramatically. This can be seen clearly in this graph of world population growth.

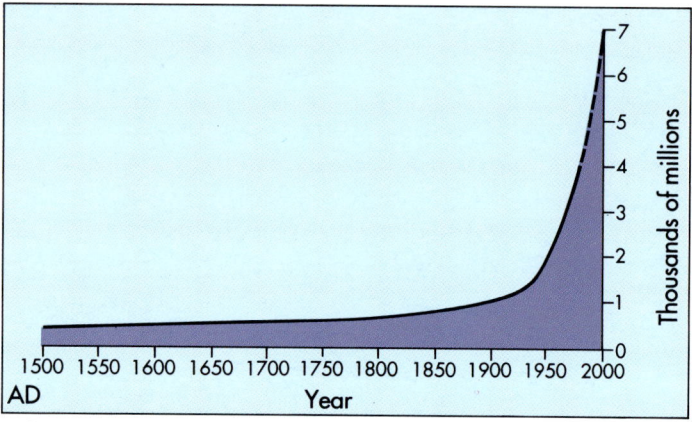

6·8 *The growth of the world's population.*

One useful way to measure the rate of world population growth is to measure the number of years the population has taken to double its size. You can see from the graph that between 1500 and 1900, the time taken for the population to double was reduced from 300 years to 100 years.

Study the graph of world population growth.

1 What was the world's population in
a 1500
b 1700
c 1900
d 1990?

2 When was the world's population
a growing very slowly
b beginning to grow more quickly
c growing very quickly?

3 What is the predicted world population for the year 2000?

4 Use the graph to make a prediction even further into the future. What could the world population be in the year of your 40th birthday?

Why has population increased so rapidly?

Case Study
B ritain

The rapid growth of population in Britain can be explained by three major 'revolutions' that started around the middle of the 18th century: in agriculture, industry and health.

Around 1750 the industrial and agricultural 'revolutions' started to take hold in Britain (and in some other European nations). New ideas were introduced initially to agriculture, such as field enclosure, and Jethro Tull's seed drill. These allowed farmers to grow food more efficiently and in greater quantities. With more and more people moving off the land to live in new industrial towns, this was a vital development.

At about the same time, the health 'revolution' began in Western Europe. In the early 18th century the death rate in England and Wales was very high – the average family had eight children, only half of whom could be expected to grow to adulthood. But there began a slow improvement in sanitary conditions, such as clean piped water and more hygienic toilets. After a cholera epidemic in 1848, in which 250,000 people died, Parliament passed the Public Health Act of 1848. A better understanding of drugs and medical needs, combined with other improvements in clothing, housing and food, led to a rapid decrease in the *death rate*. In one town, Macclesfield, for instance, the

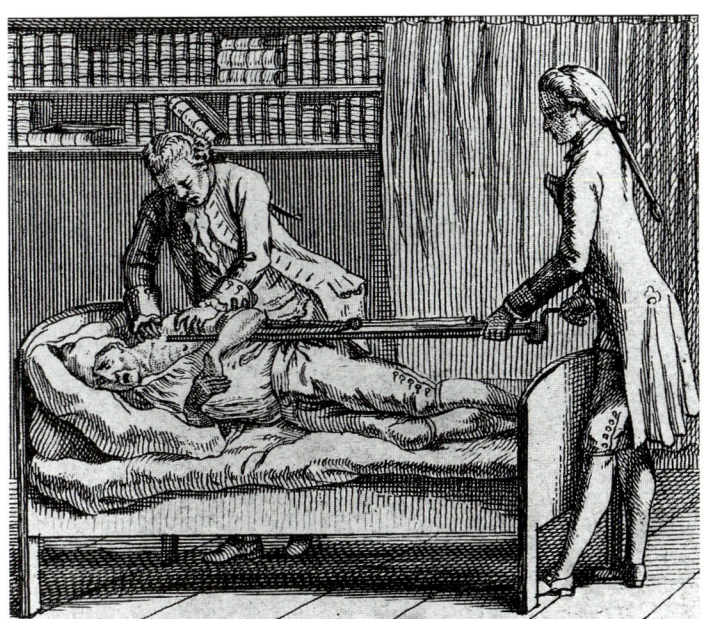

6·9 *Medical practice in the 18th century.*

death rate fell from 43 per 1,000 in 1847 to 26 per 1,000 by 1856. During this time, however, there was little change in the birth rate, which remained high.

After 1850, the death rate continued to fall, and the birth rate also began to decline. This was a period of rising living standards, and the reduced *infant mortality rate* – the death rate of very young children – meant that each child born had a good chance of surviving into adulthood. At the same time, child labour was abolished, and compulsory education introduced. Children had been a source of money, but now they were an increased cost. This was a good reason for parents to have fewer children, and so the birth rate dropped.

In the second half of the 20th century, both the birth and death rates in Britain have stabilised at low rates. The average life expectancy in Britain is now around 75, and the average number of children is about 2 per family.

These developments can be summarised in what is known as the *demographic transition model* (see next page). A model is an image of the real world that has been simplified. This makes the real world easier to study. The demographic transition model shows the stages by which a country moves from a position of high birth and death rates and low population growth (stage I), through a stage where the death rate is falling faster than the birth rate, leading to a large population increase (stages II and III), to a situation where birth and death rates are low and population growth is low again (stage IV).

Birth and death rates

Birth rate (BR)
The number of births for every 1,000 people in the country. In Britain the birth rate is about 13 per 1,000.

Death rate (DR)
The number of deaths for every 1,000 people in the country. In Britain in 1991 the death rate was 12 per 1,000.

Life expectancy
The number of years that a new baby can expect to live. In Britain on average this is 78 years for women, and 72 years for men.

Birth and death control

Death control
In Britain in the past, death was a common experience for all families and households. History shows that it is possible for society to control deaths quite easily by providing good food, water and sanitation. It is a shock sometimes to discover that nearly one-quarter of the world's people do not have these things, and are in danger of death.

Birth control
Birth control is more difficult to achieve than death control. This is mainly because it is *individual* men and women who must decide in advance that a birth should not result from their sexual relationship. If there is a good reason not to have children, *and* reliable birth control methods are available, *and* people want to use them (and know how to), then 'family planning' is possible.

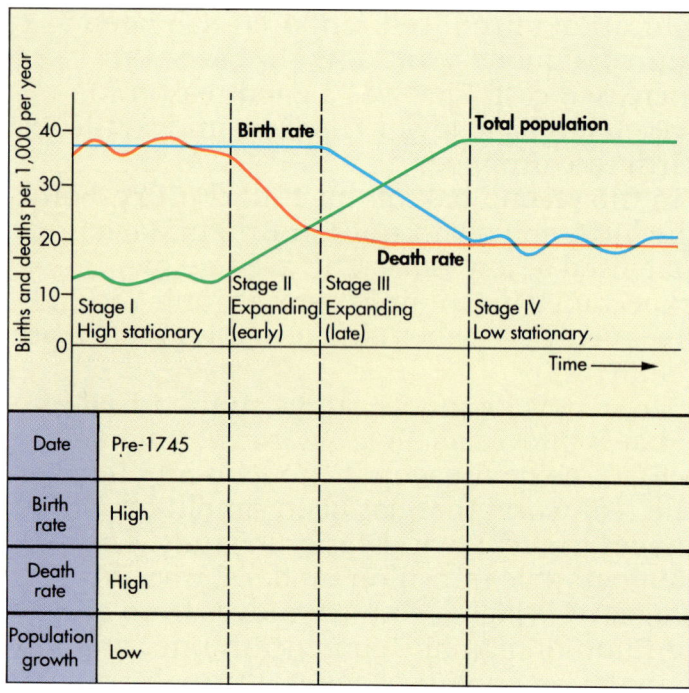

		Stage II Expanding (early)	Stage III Expanding (late)	Stage IV
Date	Pre-1745			
Birth rate	High			
Death rate	High			
Population growth	Low			

6·10   *The demographic transition model.*

You will need a copy of the demographic transition model for this exercise.

1   Using the information in the case study on pages 108–109, fill in the spaces under stages II–IV of the transition model.

2   Explain what is meant by the agricultural and health 'revolutions'.

3   Why should new methods and ideas developed at the time of these revolutions have resulted in the rapid increase of population?

4a   List some possible reasons why the birth rate fell, and has remained low, in Britain.
b   Which reason in your list do you think is the most important?

5   Do you think the demographic transition model has any 'lessons' for other countries? If so, what are these lessons?

Population change around the world

We can see then that Britain has passed through three stages of growth:

1   Low population, low growth
2   Increasing population, high growth
3   High population, low growth

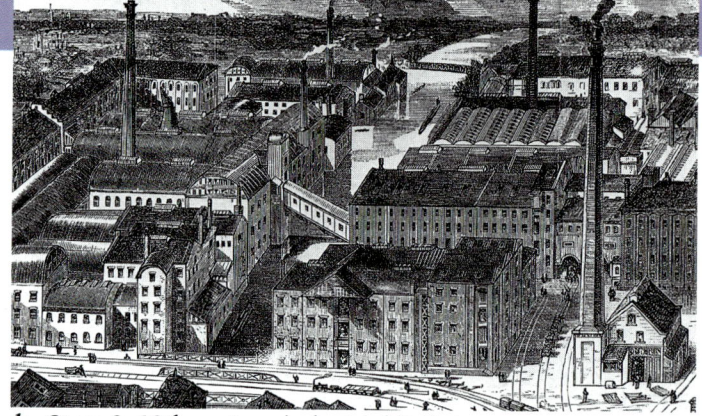

b   Stage 2: 19th-century industrial Britain.

a   Stage 1: 18th-century rural England.

6·11   *Britain's stages of growth.*

c   Stage 3: Late 20th-century urban society.

Some experts predict that Britain's population will have begun to decline by the year 2000. Yet we saw earlier that the world's population overall is growing rapidly. How can this be so?

The population of the 'majority world' – that is, the economically developing world – is still expanding fast. Individual countries of the majority world are in stages II and III of the demographic transition model. In other words, medical improvements have enabled countries to reduce the death rate; but people are still having large families and the birth rate remains high.

6·12 *World population change. The map shows that the populations of some countries are increasing rapidly, whilst others are very low or even declining.*

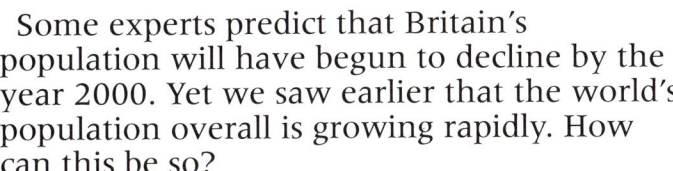

Population growth rates

Population growth rates are measured by *annual percentage change*. This can look misleading at first glance.

- Less than 1% annual change is a *low* growth rate.
- Above 3% is a *high* growth rate.
- Above 4% is *very high indeed*, though it may not sound it.

The real differences between these figures can be seen when the doubling time is calculated. This is *the number of years it takes for the population to double in size* – for example, for the population to double from 10 to 20, 100 to 200, 1,000 to 2,000, etc.

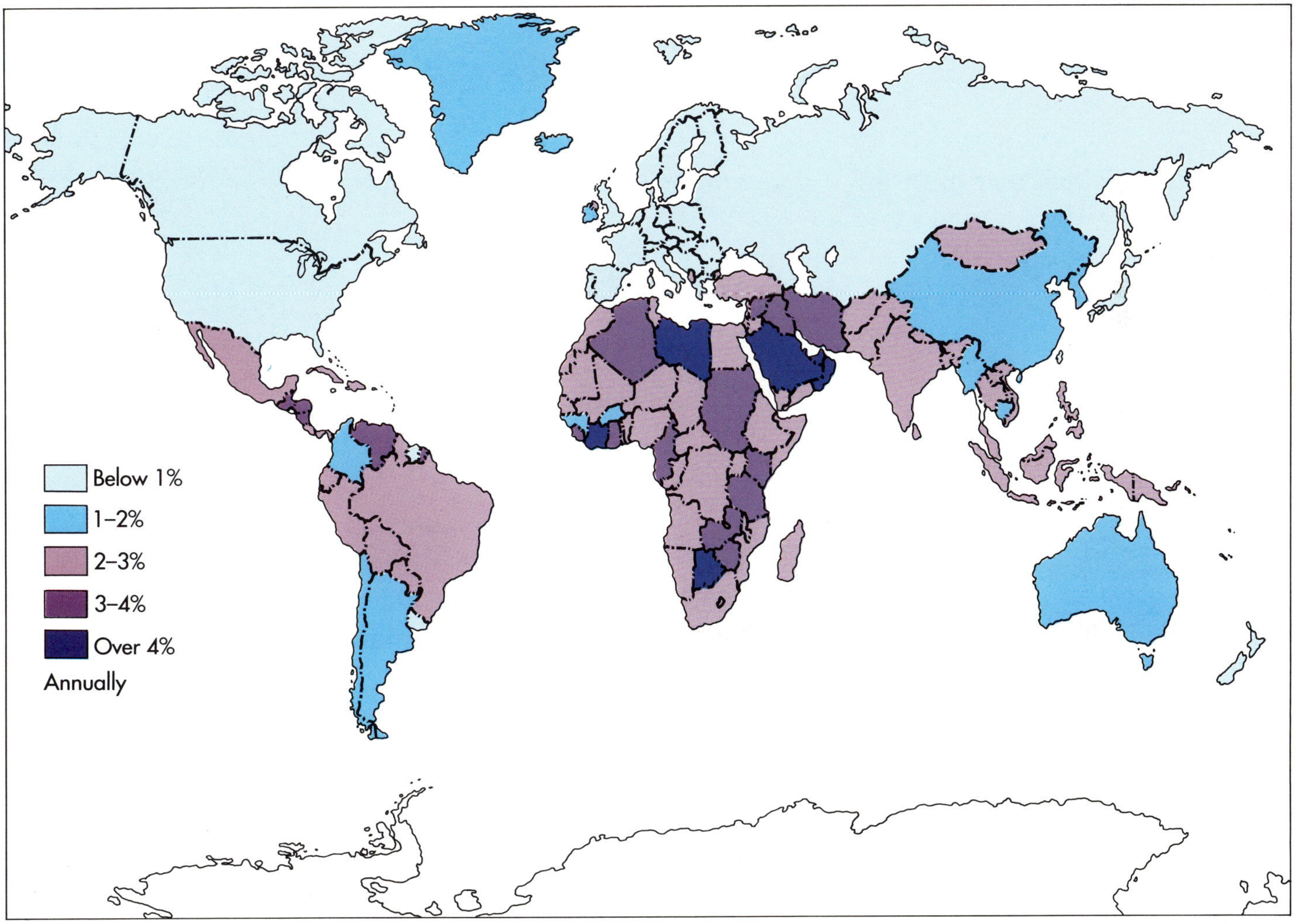

Below 1%
1–2%
2–3%
3–4%
Over 4%
Annually

Use a calculator to see the real difference between a 1% annual change and one at 4%.

1 Imagine a country that has a population of 100, and an annual growth rate of 1%.
- What is 1% of 100? (1) This is the increase in population after one year (Y1, in this case 100 + 1 = 101).
- Multiply the Y1 population by 1%, and so on until the population has doubled to 200.

How many times did you have to multiply? In other words, how many years does it take for the population to double?

2 Repeat this for another country with a population of 100, this time with a growth rate of 4%.
- What is the difference in the 'doubling time'?

Remember, if the population doubles, the country must provide double the amount of food, double the number of homes, double the number of jobs . . . and so on.

Study the map on page 111, of natural population change around the world, and refer to an atlas.

3a Name five countries that have an annual population growth of more than 4%.

b Which are the fastest-growing countries in the world?
c Name five countries in which the population is growing by less than 1%.
d Which countries appear to have a declining population?

You will need a world outline map for the next exercise.
4a Mark the following on your map: North America, South America, Africa, Europe, Middle East, Southeast Asia, China, former USSR, Oceania.
b Use the key and the information on the map on page 111 to shade in your own population change map. Use just three colours to show high, medium and low population change. Your aim is to shade whole continental areas in one appropriate colour. Give your map a title and a key.

5 Which parts of the world are growing
a fastest
b slowest?

6 Do you notice anything that the countries growing quickly have in common? Or the countries that are growing slowly?

7 Use the maps on pages 107 and 111 to identify the countries most likely to suffer the worst consequences of a population explosion; that is, countries with a high population *and* high population increase.

Rich world, poor world: population structure

Population seems to be increasing most rapidly in the poorer countries of the world, and least rapidly in the richer countries. This results in a contrast in *population structure* between rich and poor countries. In poorer countries there are many children and (relatively) fewer old people. In richer countries there are relatively few children and more old people. We saw this pattern in Unit 1, when looking at the impact of this situation as America 'grows old' (pages 18–20).

We can show population structure in a *population pyramid*. This divides up a country's population into different age categories, and then shows the proportions of males and females in each age group.

1   Which graph below represents Sweden and which represents Mexico? Explain your answers.

2   Graph A is very broad at the bottom and narrow at the top. Pick out the description from the following that is *true* of the population of that country:
- Few children and few old people.
- Many children and few old people.
- Few children and many old people.
- Many children and many old people.

3   Which country, Sweden or Mexico, has a high birth rate?

4   Which country, Sweden or Mexico, has a higher proportion of its population of a working age; that is, between the ages of 15 and 64? Why is it beneficial for a country to have a higher proportion of its population of a working age?

5   Using the information in the table below, draw the population pyramid for England and Wales in 1989. Is the pyramid you have drawn more like Sweden or Mexico?

6   What proportion of the England and Wales population is over 60?

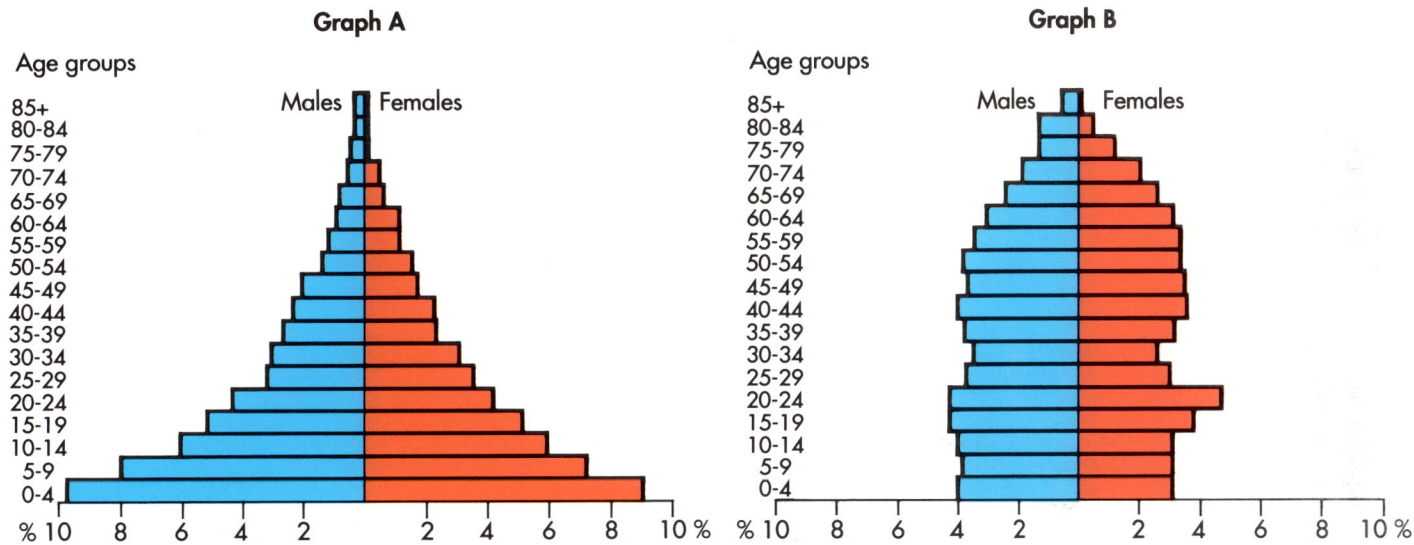

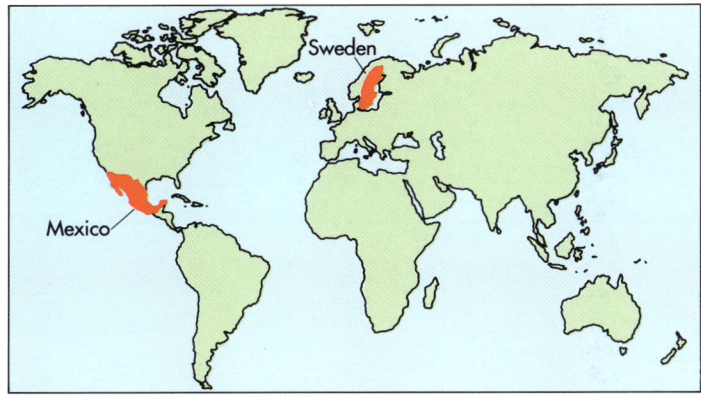

6·13   *Population pyramids for Sweden (a richer country) and Mexico (a poorer country).*

6·14   *The age and sex structure of England and Wales, 1989.*

Age group (years)	Percentage of total population	
	Males	Females
0–9	6.6	6.3
10–19	6.6	6.4
20–29	8.2	8.0
30–39	6.8	6.7
40–49	6.5	6.4
50–59	5.2	5.3
60–69	4.8	5.4
70–79	2.7	4.1
80–84	0.7	1.4
85 and over	0.4	1.1

▷ The seesaw: How can the world balance population and resources?

Having children is a matter between individual men and women. On a world scale their decisions can add to the population. Each new human needs food, shelter, water, and many other things besides. So each additional person uses a little more of the Earth's resources.

A small number of people and a large quantity of natural resources might be drawn like this on a balance (weighing scales).

A larger number of people with those same natural resources might be shown like this. The population and the resources are in balance.

A growing population will have a greater need for natural resources. If those resources are not managed properly then there will be too few resources for the growing population. Then we are in trouble.

6·15 *Sitting on the seesaw.*

Look at the interviews on pages 115–116 with people around the world. There are five families to meet.

1 Work in small groups. In each group, one person (or two people) should take the role of the family member/members being interviewed. Another person should be the interviewer. Others in your group should listen, and work out the reasons for the size of the family in each case. Remember to rotate the roles.

2 People around the world have different ideas about the size of their family.
a Make a list of the different reasons why people decide on the size of their family.
b In your group, try to work out if any of these reasons are specific to countries in the southern or northern hemisphere. Or are some reasons common to both? You could present the information in two overlapping circles, like those you used earlier on page 104.

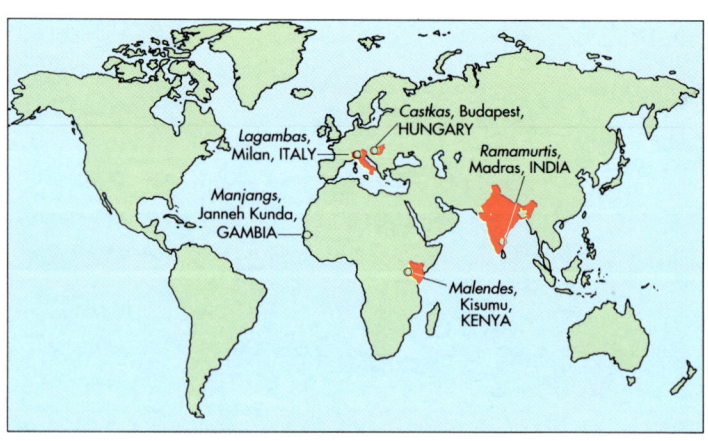

Silvana Lagamba from Milan in ITALY

Our friends' experience is the same as ours. They don't even have two children like us, often not even one! Sometimes they even give up starting a family. It doesn't only apply to our circle of professional people – it's quite a common phenomenon in all social classes in Italy. Family size has gone down a lot.

In part, this is due to women's role in society changing. Women are now involved in the working world so they have less time to look after the family and children. Then when you take into account the lack of social services, it's only possible to have a few children. Even now there are times when we feel a sort of rejection towards our son. Well not quite, but a kind of rejection since deep down we see him as a sort of competition. In other words, the fact that we have to look after him and his needs, we sometimes feel it is an alternative to other needs that we might have. We haven't managed to achieve a balance between our own needs as individuals and as parents. I'm not sure we'll ever achieve this.

Mr & Mrs Ramamurti from Madras in INDIA

In the past we were poor. Even now we're not rich. We've managed to accumulate some material wealth. After all our hard work there should be a child to enjoy it. Then we too could enjoy it.

In our way of life, according to our caste, there should be a son to perform all the religious rituals on his father's death. I don't have a son to do that. That's why I'm starting to feel bad now.

My neighbour's children climb on top of me calling "Uncle, uncle!" I don't mind my rest being interrupted. Playing with the children makes me happier than sleeping, but it also reminds me of our regret at not having children. We were eagerly expecting until the age of 50. "Will it happen? Will it be born?" When we passed 50 we became weak in our health. Frankly there's no enjoyment in life. How can we expect to have children now?

It's the happiness of children we want. Money, we can earn. We can't buy the happiness of children.

Florence Malende from Kisumu in KENYA

My family was a family of girls. That is a problem – having one sex. Here in Kenya they believe that if a man doesn't have a boy then you are not a man. You cannot sit with other men and discuss their problems.

Florence Malende is a health worker. She has five children. The eldest three are girls, the youngest two are boys.

Mrs & Mr Castka from Budapest in HUNGARY

We've been thinking about having a second child for 2½ years. Julia, our daughter, would like one, but we haven't made our decision yet.

There's a lot to think about, having a second child. There's the question of money, there's our other commitments. We're not sure we should take on the burden of a second child.

If our flat was bigger there'd be room for two children. We have one room and two half rooms. There would just be room for another child but the loan repayments on the apartment are high already, and then I think a second child would get in the way of my work. Nowadays you have to think of your career. If I had a second child it would be the end of my career. First of all you're pregnant for nine months. My first pregnancy was when I was a student so there wasn't a problem. Then there's another year before the child grows big enough for you to have some freedom. It's two years out of a woman's life.

Both of us have to want it. I can't do it on my own! We'll decide in a civilised manner whether we do or don't. In the end I'm against it mainly because my wife doesn't want it.

Most people these days work hard to make money. Teachers like me give extra lessons, but I spend my spare time with Julia. If I had two children I'd have to give more lessons so that I could provide as much for the two as I do for the one. Then the family isn't living like a family any more.

Isatou Manjang from Janneh Kunda in
THE GAMBIA

I have spent the whole of my life in Janneh Kunda, as have my parents. My father has 15 surviving children - I can't say exactly how many died - it was 6 or 7, I think. I am one of two daughters and three sons of my father's second marriage. Like my parents, I didn't go to school. I spent my childhood helping my mother at home or on the land. I learnt a lot from my mother. We have a small vegetable garden with a well. I have 8 plots on which I grow onions, tomatoes, peppers, aubergines and cabbages, all of which we eat in soups and stews. If there is any produce to spare, I sell it and use the money to buy other foods or clothes.

I was married a few months ago, but I still spend most of my time at my parents' compound. I only visit my husband's compound occasionally. Only after two or three years will I start to live with my husband permanently. When we have children I want whatever God gives me! I will of course look to my husband to make all the major decisions. He will want many children. He is older than me. I am his second wife. I want many children too - to help me with the daily work, and maybe they will go to our new school.

Population and resources

The number of children that people have affects the world's resources, and has an impact on the environment. What images do these two illustrations present about population, resources and the environment?

In Unit 8 of *Green Pieces* we examined the relationship between people and the Earth's resources. It is an issue that concerns the whole world, and seems to become a greater problem with every year that passes.

Source: State of World Population Report, UNFPA.

6·16 *If the world's largest cities continue to grow at current rates, then health and sanitation systems could be overwhelmed.*

6·17 *As human numbers increase, the resources on which we depend – land, water and air – will come under such unbearable pressure that our planet could become uninhabitable.*

THE INDY

YOUNG NEWSPAPER PUBLISHING LTD

The rapid and continuing rise in population is placing great strain on the world's resources. Lucy Dalrymple reports

Bursting at the seams

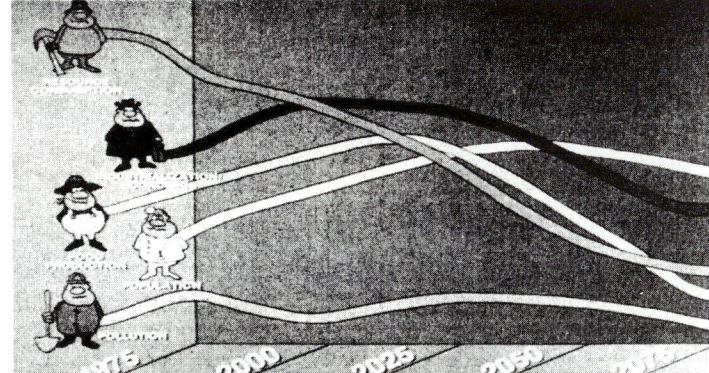

The rapid growth of the world's population is straining the earth's resources

"THE POPULATION of the human race is now 5.3 billion, and still growing. In the six seconds it takes you to read this sentence, 18 more people will have been added to the world."

This shocking fact appears in a recently published book *The Population Explosion* by Paul and Anne Ehrlich. When these authors first warned of the consequences of rapid population growth on the environment many regarded them as cranks.

Now some of the statistics are horrifying. The 1989 World Population Data Sheet shows that if the increase continues at the current rate, the world's population will double in 39 years. How could the world to cope with 10 billion people?

Rich countries, including the UK, are straining the world's resources with our enormous food and consumer goods consumption.

The birth of a baby in a rich country is a greater threat to earth's life support systems than that of a baby in a poor nation.

The average child from a rich country consumes between 20-30 times more natural resources than a child in Bangladesh. This has a negative impact on food and water supplies.

At the United Nations Conference on World Population in Mexico in 1984 all 130 nations represented agreed that something should be done.

Eric Deakins of Population Concern comments: "Governments and individuals must take action now.

Rapid population growth in poor countries undermines any efforts to alleviate poverty".

John Rowley of the International Planned Parenthood Federation believes efforts to slow growth must come with education and health care.

He believes many measures could be implemented quickly. Energy conservation should top the list, and the West should be encouraged to share its conservation technologies with other, less advanced countries.

But most important is the spread of information on family planning in developing countries. In many countries the burden of childbearing can shorten life.

Worldwide every year, half a million women worldwide die from pregnancy-related causes, 96 per cent of them in poor countries.

By improving the social position of women through education they can choose when to have children. By promoting choice through education many women can be spared the threat of losing their job, incurring fines or being forced to have an abortion if they do become pregnant.

Further information

For more information about world population contact:
* *Population Concern, 231 Tottenham Court Road, London W1F 9AE*
* *International Planned Parenthood Federation, Regent's College, Inner Circle, Regent's Park, London NW1 4NS*
* *Committee on Population and the Economy, 13 Norfolk House, Courtlands, Sheen Road, Richmond, Surrey TW10 5AT*

6·18 *From The Indy, 24 May 1990.*

Read the newspaper article on page 117.

1 According to the writer, in which part of the world is the birth of a baby a greater threat to the Earth's resources?

2 Why is rapid population growth such a problem to poorer countries?

3 What effect do you think each of the following could have on controlling the world's population growth?

- Getting rid of poverty
- Improving the social position of women
- Making sure that all children around the world are educated (learn to read and write)

4 Find some newspaper reports on the subject of population. Display these on the classroom noticeboard.

Migration

High population growth rates in poorer countries give us one reason why these countries remain in poverty.

When people feel they are condemned to live in poverty, they often choose to move away, to save themselves and to seek a better life for their children. This has happened all through history. The USA, for example, has received *immigrants* from all over the world. In 1907 alone, 1,285,000 Europeans entered the USA. Today, people from Central American countries like Mexico make up most of the immigrants. Some experts predict that in the future the majority of US citizens will be Spanish speaking.

Some *migrants* are forced to leave their homes for other reasons. Many of the Europeans who moved to the USA earlier in the 20th century were Jews in fear of persecution. There are still many people around the world who are in fear of their lives. Countries of the EC have taken in such *political refugees* from many countries, including Chile, China, Croatia, Iran, Somalia, Sri Lanka . . . the list is a long one.

The Statue of Liberty, New York.

6.19 'The American Dream'.

YOUNG NEWSPAPER PUBLISHING LTD

The call of the American dream

Give me your tired, your poor.
Your huddled masses, yearning to breathe free.
The wretched refuse of your teeming shore.
Send these the homeless tempest-tost to me.

SO SAYS the inscription on the Statue of Liberty. But this was written in 1883: a great deal has changed since then. America's shores are now just as teeming as anyone else's. Illegal immigration is a very serious problem.

Countless hopefuls run the gauntlet of US border patrols every year. According to the Immigration and Naturalisation Service (INS) about three million make it through.

About half of this figure are aided by professional "people-smugglers". It is a booming business, netting an annual tax-free profit of $1 billion.

The smugglers' efficiency is startling: they provide passports, visas, safe houses and bribes. They smuggle people from anywhere and everywhere; but at a price. This is mainly worked out by these factors:
How far the alien has to travel.
How difficult it is to get them out of their country (not easy in China).
How much money they have got.

At the top end of the market, the service is almost luxurious. In April, 47 Chinese were nabbed as they stepped off a plane in Montreal, Canada. They had flown first class.

The other end of the scale is far more sordid. In 1987, smugglers locked 19 Mexicans in an airtight steel railway carriage in El Paso, Texas. The temperature rose to 130 degrees Fahrenheit. When they were found several hours later, all but one had suffocated.

Your huddled masses, yearning to breathe free suddenly takes on rather a terrible irony.

From The Indy, 24 May 1990.

There are therefore two types of migrant:
- the *economic migrant* and
- the *political refugee*.

In reality the economic migrant feels like a refugee too, the choice being between migration (and hope) and misery (including possible starvation).

Richer countries, especially those in Europe and North America, are fearful of the possible effects of large-scale migration. Barriers and controls are being created to keep migrants out. The newspaper article opposite reports on the results of such a 'fortress' policy.

The aim of the next exercise is to link the model of migration to a real example of migration that you know about. This model is known as a *push-pull model*. What it shows is that a person who chooses to migrate from one country to another does so for two sets of reasons: 'push' reasons and 'pull' reasons.

1 In pairs, study the model carefully. Describe what you think it says, to make sure you understand it.

2 Choose an example of *international migration*, either from this unit or from another source that you know about. On a large piece of paper, draw a model of your example, using as much exact detail as possible. The model on this page is a *general model*. Yours will be a *specific model*.

6·20 *Model of migration.*

KEY ▶ POINTS KEY ! IDEAS

Natural changes of population

People are born and people die. The difference between the number of births and deaths gives the *natural change* of the population. This can be shown as an equation:

$P_{(N)} = B - D$

where

- $P_{(N)}$ = natural change in population
- B = number of births
- D = number of deaths

If deaths outnumber births, the population decreases and the natural change is *negative*.

Migration

Migration is the other reason why populations change. The equation becomes:

$P = B - D \pm M$

where

- P = population change
- B = number of births
- D = number of deaths
- M = number of migrants

A migrant is a person who moves to another place to live, permanently. There are many reasons why people choose to do this. For millions of people the reasons are economic – they become *economic migrants* in search of jobs or better opportunities. For millions, the reasons are political – people from all parts of the world at various points in history have become *refugees*, often in fear of their lives.

Demographic transition

A *demographer* is someone who makes a special study of population. Many demographers believe that the population of

countries must pass through certain stages before the world's population is brought under control. Countries must change from having a high birth rate and high death rate to having a low birth rate and low death rate.

Population distribution

The 5 billion (or thereabouts) people of the world are not spread evenly across the land surface of the Earth. There is a population distribution pattern. The pattern is not just an accident, and can be explained in various ways: to do with the physical world (how this can encourage or discourage people), and the human world (how people can use their skills and brains to overcome physical obstacles).

Population structure

The population of a country is counted by a *census*. The census provides lots of other details which can be used to compare populations. One useful aspect is to compare the age-sex structure of population. Studying the structure of population helps a society plan for the future: how many schools will be needed, how many hospitals, retirement homes, and so on.

Population and resources

Populations make demands on water, land and air. People need food and fuel, homes and workplaces, and they produce waste and rubbish which need to be disposed of. The wealthier the people, the greater the demands they make.

Living a Life of Leisure

7·1 *Signs of tourism among the coconut palms. The tourist industry's resources include the world's quiet places. 'Paradise', though, is constantly under threat.*

Brainstorm: Holidays around the world

Arrange yourselves into groups.

1a Make a list of all the places you can think of that are popular holiday locations.
b Now try to classify the places on your list. To do this, organise the places under a number of different headings to show the different types of holiday, for example:

c Write down some reasons why you think people like to take the types of holidays you have identified.

2 Now look again at the photograph on page 121. What changes happen to the people and places that receive tourists?

Activity	Sun resort	Winter resort	Culture

Setting the scene

What sorts of holiday places did your group identify? How does all this tourist activity affect people's lives? And how will tourism develop in the future?

Modern life can be full of pressures and responsibilities. It is no surprise then that people like to make the most of their free time. Taking a trip or a holiday, 'getting away from it all', is one way to relax.

Think about when you have leisure time. It is basically all the time when you are not eating, sleeping, attending school or doing homework and household chores. That leaves weekday evenings, weekends and holidays. You can plan your leisure activities accordingly: perhaps go to a club in the evenings, have a day out at the weekends, and a trip away in the holidays.

Travelling away from home for pleasure is now part of our way of life.

7·2 *The fantasy world of Disney has become a reality for European tourists.*

Leisure society

Ask an elderly relative or neighbour about their holiday experiences as a child. They will probably tell you that holidays were rare luxuries. Foreign holidays were virtually unheard of. Nowadays more and more people have the *time, money* and *mobility* to be able to spend part of their leisure time away from home. In the first half of the 20th century few workers were able to afford to take holidays at all because they lost their pay for the time they were away. Today holidays are longer and workers are paid during their holiday. The increased mechanisation of industry has reduced the length of the working week too.

The average working wage has risen to give people more *disposable income* – money that can be spent on non-essential items.

Elderly people have a lot of leisure time, and the number of people over the age of 65 is rising rapidly. More and more people are retiring early so that they can enjoy their leisure time while they are still fit and healthy.

Alongside these changes there have been advances in transport technology, making people more mobile. In 1951 just 1 in 20 Britons owned a car; by 1990 there was an *average* of almost one vehicle per family. There have also been huge advances in air travel in the latter part of the 20th century. Air travel has become cheaper, faster and more comfortable on new jet planes.

An outcome of these developments is that we now live in a *leisure society*.

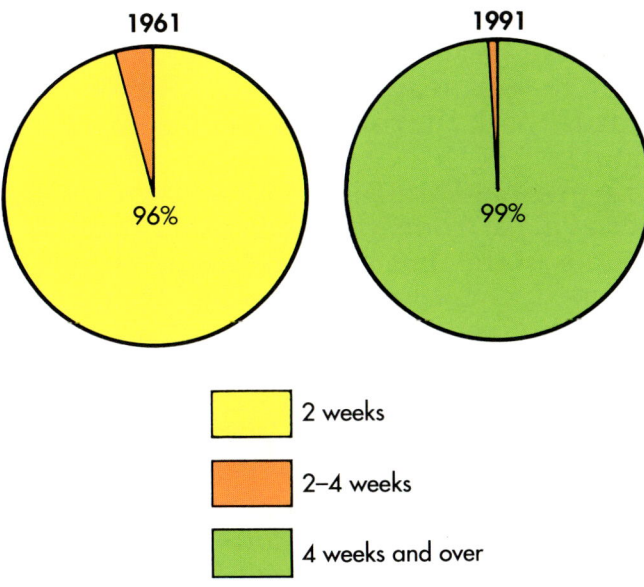

a Holidays with pay for British employees

- 2 weeks
- 2–4 weeks
- 4 weeks and over

7·3 *The leisure society is characterised by increased levels of free time, money and mobility.*

Leisure society

More people than ever before are able to take part in activities for pleasure. We have
- more free time
- more disposable income, and
- greater mobility.

b Average weekly earnings of full-time working adults in Britain

	Men	Women
1970	£27.20	£14.60
1980	£113.30	£72.40
1990	£258.20	£177.50

Sources: *Annual Abstract of Statistics, 1972, 1982, 1992, Central Office of Statistics, HMSO, London.*

c Charter passengers handled by London's Gatwick Airport, 1960–90.

Year	Passengers
1960	140,000
1965	850,000
1970	2,400,000
1975	3,600,000
1980	5,500,000
1985	8,400,000
1990	8,000,000

Source: *BAA Airports Traffic Statistics, 1990/91, Research Dept, BAA plc, June 1991.*

1 Explain what you think is meant by the expression 'getting away from it all'. Make a list of the things people might want to get away *from*.

2 Now make a list of the features that might attract people to visit the EuroDisney pleasure park.

3 Study your two lists carefully. What differences do you notice between them?

4 Using data from the tables above, explain the changes that have led to the leisure society.

The tourism explosion

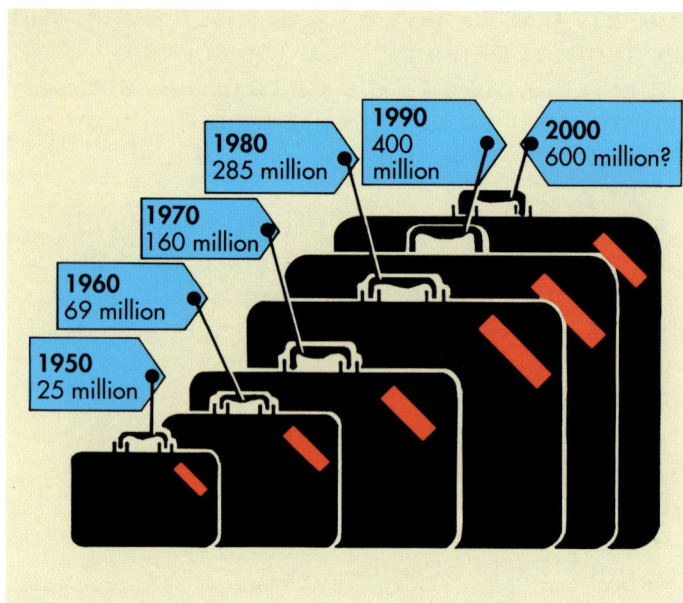

Park	Annual attendance
Walt Disney World/Magic Kingdom/ Epcot, Florida, USA	25,100,000
Disneyland, California, USA	13,000,000
Pier 39, San Francisco, California, USA	10,465,000
Blackpool Pleasure Beach, UK	6,500,000
Sea World of Florida, USA	4,584,000
Universal Studios Tour, California, USA	4,240,000
Tivoli Gardens, Denmark	4,000,000
Knott's Berry Farm, California, USA	4,000,000
Busch Gardens, The Dark Continent, Florida, USA	3,720,000
Sea World of California, USA	3,350,000

From The Observer, 1 March 1992

7·5 *The world's top 10 tourist attractions.*

7·4 *There has been a 1,600% increase in the number of international tourists between 1950 and 1990. An explosion in tourism!*

The growth of the leisure society has resulted in a tourism explosion.

In a very short time tourism has generated 100 million jobs throughout the world, which means that 1 in 5 of the world's workers is employed in a tourist-related job. It is estimated that by the year 2000 the tourist industry will be the most important economic activity in the world, making more money than even the oil industry. Tourism is seen by many developing countries as a way to develop the economy, by creating jobs and bringing in money from foreign visitors.

Tourism

People who travel to an area away from home in their leisure time are known as *tourists*.

International tourists are tourists who travel to a different country.

The services set up to cater for the needs of tourists form the *tourist industry*.

And where do people like to go to enjoy their leisure time? The table above indicates that there has been a 'Disneyfication' of our desire to 'get away from it all'. Fantasy pleasure parks have become enormously popular.

The original Walt Disney idea was to create a special place from which the unpleasant realities of the outside world could be excluded. There is no litter, no graffiti, no bad behaviour – everything is neat, clean and well ordered. Once inside a pleasure park, visitors experience a sense of 'placelessness' – in other words, visitors to EuroDisney in France have little way of telling which country they are in. Placelessness is also a feature of large modern shopping malls. And is it reassuring or worrying that Macdonalds is almost identical *wherever* in the world you order your burger? The popularity of pleasure parks is in a way a reaction to the impact that the leisure society and tourism have had on the world's *real* societies and places.

In this unit we study some of these impacts and answer three key questions:

▷ Are national parks just big public playgrounds?

▷ How much do holidays really cost?

▷ Is tourism a passport to development?

▷ Are national parks just big public playgrounds?

In the USA, roads to the world's oldest national park have 'Yellowstone Full' signs displayed 150 kilometres from its boundary on busy days.

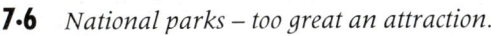

<u>90</u> MILLION VISITS ARE MADE TO THE 12 national parks of England and Wales every year. Some people believe that in the future we will have to book in advance for a trip to the most popular national parks.

7·6 *National parks – too great an attraction.*

Read the statements above. A national park sounds like a big playground for the public to use in their leisure time. This is only partly what national parks are about. In fact, national parks have been created in response to a *variety* of pressures felt by society and the environment.

The idea of national parks dates back to 1872 when the US government gave the wild countryside of Yellowstone in the Rocky Mountains special protection by law. At around the same time there was an upsurge of pressure in Britain for better access to wild and open country for the millions of people living in industrial towns and cities. Finally, in 1950, the National Parks and Access to the Countryside Act was passed.

7·7 *A rambler from Manchester sings the words of a popular folk song.*

From 'National Parks in England and Wales', Countryside Commission, 1986

7·8 *The aims of the National Parks Act.*

Under the Act a total of 12 national parks have been created in England and Wales. They cover about 10% of the total land area of the two countries.

Robin Hood's Bay, North York Moors.

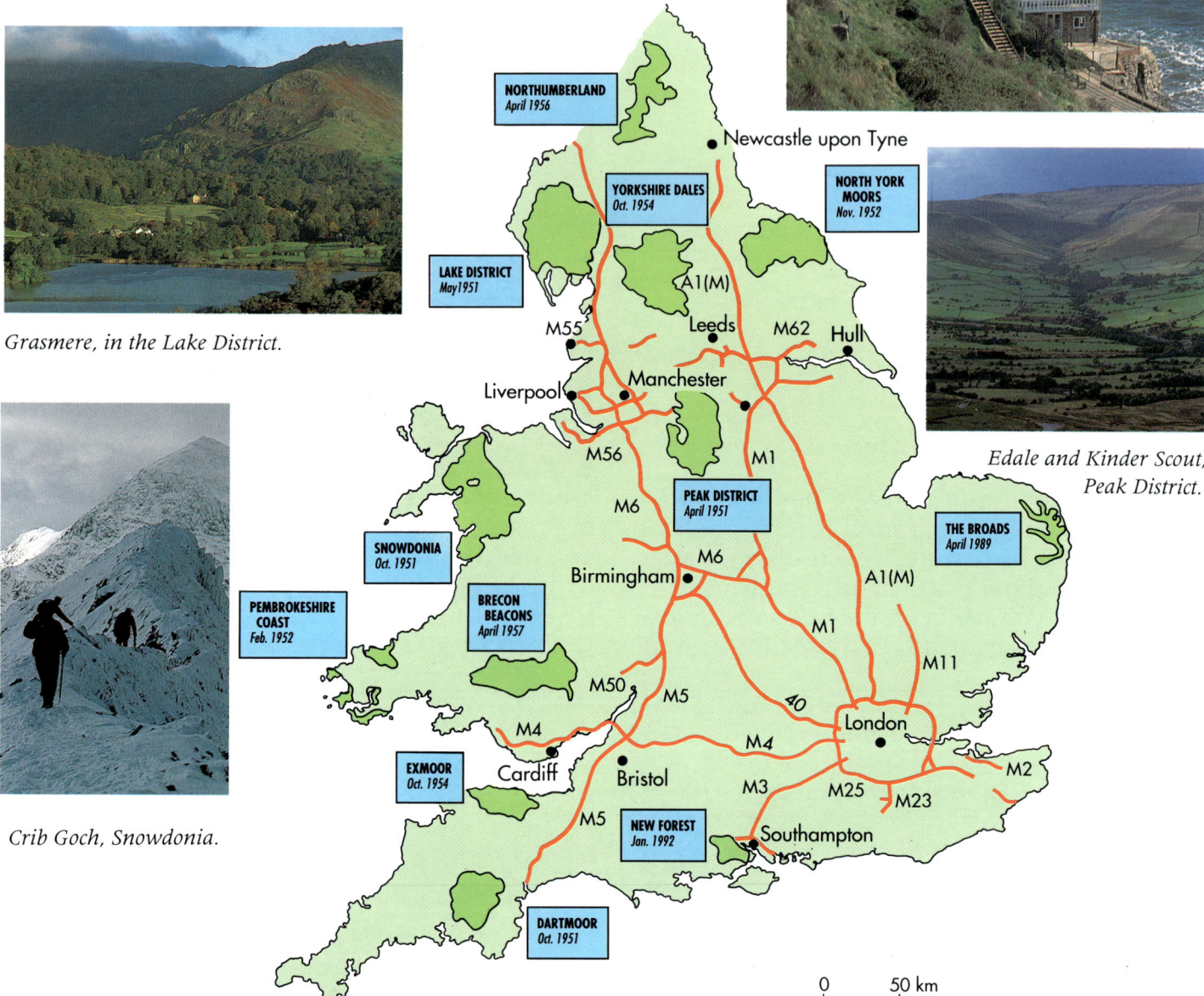

Grasmere, in the Lake District.

Crib Goch, Snowdonia.

Edale and Kinder Scout, Peak District.

NORTHUMBERLAND
April 1956

Newcastle upon Tyne

YORKSHIRE DALES
Oct. 1954

NORTH YORK MOORS
Nov. 1952

LAKE DISTRICT
May 1951

A1(M)

M55   Leeds   M62   Hull

Liverpool   Manchester

M56

M6

PEAK DISTRICT
April 1951

M1

SNOWDONIA
Oct. 1951

THE BROADS
April 1989

BRECON BEACONS
April 1957

M6

Birmingham   A1(M)

PEMBROKESHIRE COAST
Feb. 1952

M1

M50   M5   M11

M4

EXMOOR
Oct. 1954

Cardiff   Bristol   40   London

M4   M3   M25   M23   M2

M5

NEW FOREST
Jan. 1992

Southampton

DARTMOOR
Oct. 1951

0   50 km

7·9 *The 12 national parks of England and Wales: where they are, and when they were created.*

1 In your own words, list the aims of national parks as stated in the John Dower Report.

2a Using the map above, list the names of the 12 national parks.
b In which decade were most of the national parks created?

c Which two have been created most recently, and when?

3a Which national park is closest to where you live?
b Describe how you would get there if you were to plan a visit.

What are the main features of national parks?

National parks try to accommodate both *conservation* and the *leisure needs* of society. They have a special character. What are the main features of national parks?

Key

–·–·– Boundary of national park	◙ Accessible water site
Relatively high ground	v Viewpoint
Motorway	▲ Youth Hostel
'A' class road	c Camping/caravan site
----- 'B' class road	*i* Tourist Information Centre
● Town/village	⊕ Mountain Rescue Post

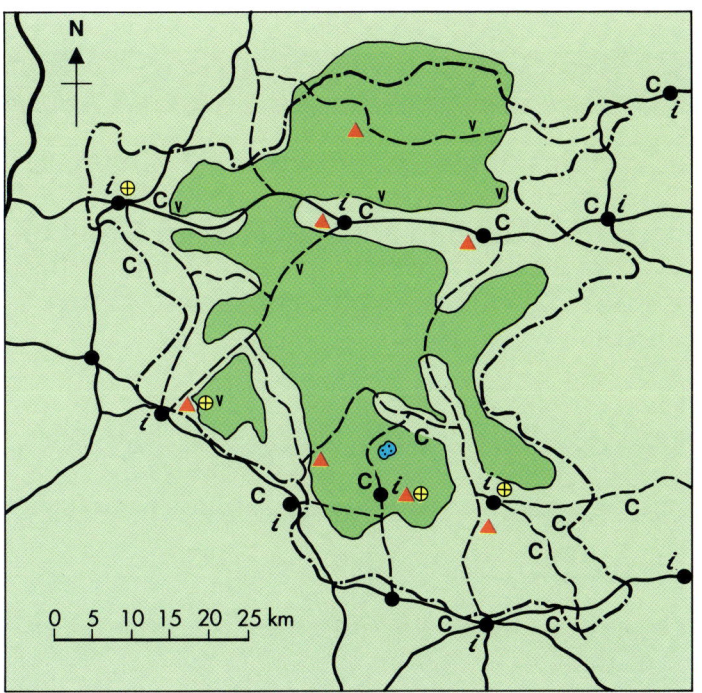

7·10 *The Yorkshire Dales National Park.*

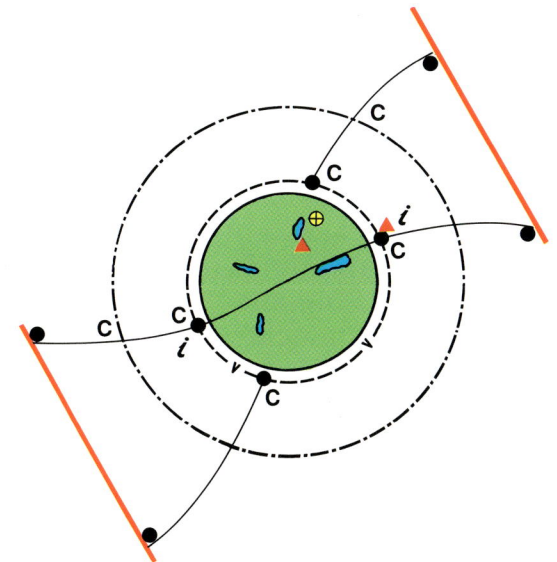

7·11 *A British national park model.*

NAME OF NATIONAL PARK _____				Just outside national park	On periphery of national park	In centre of national park
1 PHYSICAL FACTOR	Key	Yes	No			
a Is the national park an upland area compared with the surrounding country?	(high ground)					
b Is the national park a compact unit?	–·–·–					
c Are watersites available?	◙					
2 HUMAN FACTORS	—					
a1 Is the national park served by motorway?						
a2 Is the national park served by 'A' class roads?	—					
a3 Is the national park internally connected by 'B' class roads?	-----					
b1 Are there camping/caravan sites?	c					
b2 Are there Youth Hostels?	▲					
c1 Are there Mountain Rescue Stations?	⊕					
c2 Are there any viewpoints?	v					
c3 Are there any Information Centres?	*i*					

7·12 *A matrix for testing the national park model.*

Source: R.T. Goring (1977) 'A British National Model', in *SAGT* No. 6.

To find out what the main features of national parks are we can use a model of British national parks. Look back to page 109 to find out why we use models. In this exercise you will 'test' the model of a national park to see how good an image of the real thing it is.

1 Use the matrix on page 127 and compare the map of the Yorkshire Dales National Park with the model.

2 How accurate an image of the Yorkshire Dales National Park is provided by the model?

The Yorkshire Dales National Park

The Yorkshire Dales National Park is visited by 8 million people every year and has 15 million urban dwellers living within one hour's drive of its boundary.

But the Dales are not only a leisure amenity, they are also important in other ways. More than 50% of the parkland is open moor which supports much plant and animal life. Some of the 18,600 people who live in villages in the park itself are farmers. Over 40% of the land in the park is used for agriculture (mainly for stock and dairy farming).

In some villages tourism has altered the traditional way of life. Visitors provide a new source of income, and new industries have been set up. The village of Malham, for example, is close to the popular natural attractions of Malham Cove and Malham Tarn, near the southern boundary of the park. The Malham area has about 0.5 million visitors a year. A very popular place like this is known as a 'honeypot' because so many visitors swarm to it.

1 When do you think Malham is at its busiest? Do you think most visitors are day-trippers or tourists on longer holidays?

2 Imagine you are a day visitor to Malham. Use the map opposite to plan a two-hour walk starting and finishing at the car park.

3a Footpath erosion has been identified as a problem where there are large numbers of visitors. What other problems do you think might arise in Malham with 0.5 million visitors each year?
b Devise a plan to reduce these problems in the Malham area.

4 Use the information in this section to write a magazine article about leisure in the Dales. You should include the following:
- A headline.
- The features of the Yorkshire Dales National Park and the types of people who might enjoy them.
- Brief interviews with a local trader and a conservationist giving their views of tourism in the Malham area.
- A conclusion explaining why you recommend or don't recommend a trip to the National Park.

7·13 *The view from the top of Malham Cove looking south towards Malham village and the Yorkshire Dales National Park.*

THE THREE PEAKS

of the Yorkshire Dales National Park – described as suffering from the worst erosion in Britain – are to be saved in a rescue operation costing hundreds of thousands of pounds. Most of the paths over this popular attraction for ramblers and tourists have been trampled to an average width of 11.4 metres – more than twice the size of a 'B'-class road.

by CHRIS WOOD
Yorkshire Dales National Park

From 'Paths as Wide as Roads' in 'National Parks Today 15', Autumn 1986

MOUNTAIN BIKE HIRE

EXPERIENCE THE YORKSHIRE DALES CYCLEWAY.

Hire a bike at reasonable rates per hour.

Contact the Malham Youth Hostel, open everyday May–October, bank holidays and weekends out of season.

Working Sheepdog Demonstrations

All weekend afternoons from June to September. Spend a few minutes or a few hours watching local farmers and their dogs.

In the large field behind the Malham Information Centre.

The Malham Tea Rooms

*Lunch and cream teas served allyear round.
Next door to the Buck Inn Public House.*

Malham Arts and Crafts Shop

Pottery and paintings
Open from May to September
(opposite the Post Office)

7·14 *The Yorkshire Dales National Park: the problems of popularity.*

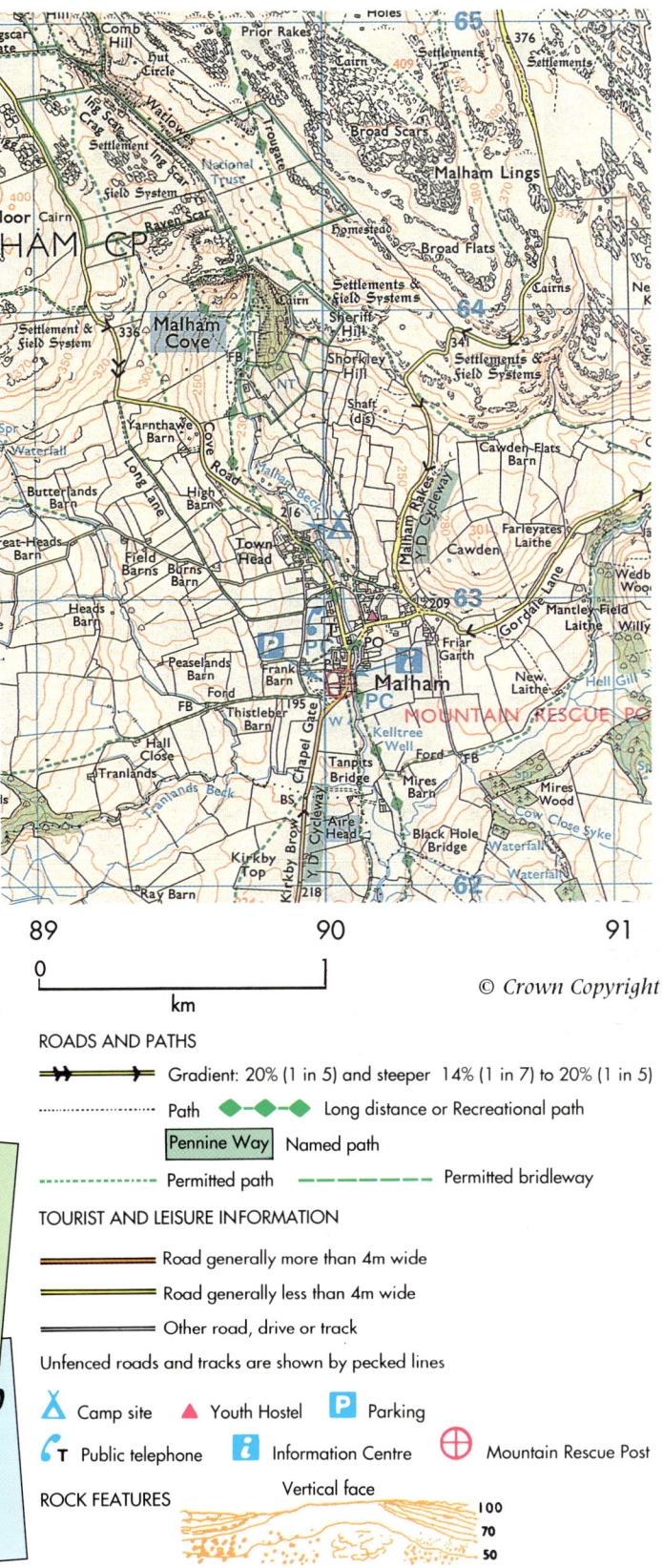

© Crown Copyright

ROADS AND PATHS

Gradient: 20% (1 in 5) and steeper 14% (1 in 7) to 20% (1 in 5)
Path Long distance or Recreational path
Pennine Way Named path
Permitted path Permitted bridleway

TOURIST AND LEISURE INFORMATION

Road generally more than 4m wide
Road generally less than 4m wide
Other road, drive or track

Unfenced roads and tracks are shown by pecked lines

Camp site Youth Hostel Parking
Public telephone Information Centre Mountain Rescue Post

ROCK FEATURES

Vertical face 100 70 50
Loose rock Boulders Outcrop Scree

From OS 1:25,000 Outdoor Leisure Sheet 10, 1989

▶ How much do holidays really cost?

In Britain in the depth of winter, TV commercials, glossy brochures·and newspaper adverts begin to offer us tempting visions of sun-tans, glamour, health and luxury in the form of foreign holidays. The prospect of 'getting away from it all' lures millions of us into accepting the invitation.

We have our desires catered for by travel companies that neatly organise holidays into ready-made 'packages'. The two most popular types of holiday are sun, sea and sand holidays, and mountain skiing holidays.

The world's most popular sun, sea and sand holiday destination is the Mediterranean coast of Europe. It is estimated that in 1985 there were over 100 million visitors to the Mediterranean. The world's most popular mountain region is the European Alps where 40 million tourists spend their holidays each year.

In 1987 nearly 27.5 million British tourists holidayed abroad; 80% of those holidays were in European countries.

a Benidorm on the Costa del Sol is a popular sun, sea and sand resort on Spain's Mediterranean coast.

b Livigno in the Italian Alps attracts skiers from all over Europe.

7·15 *Popular European resorts.*

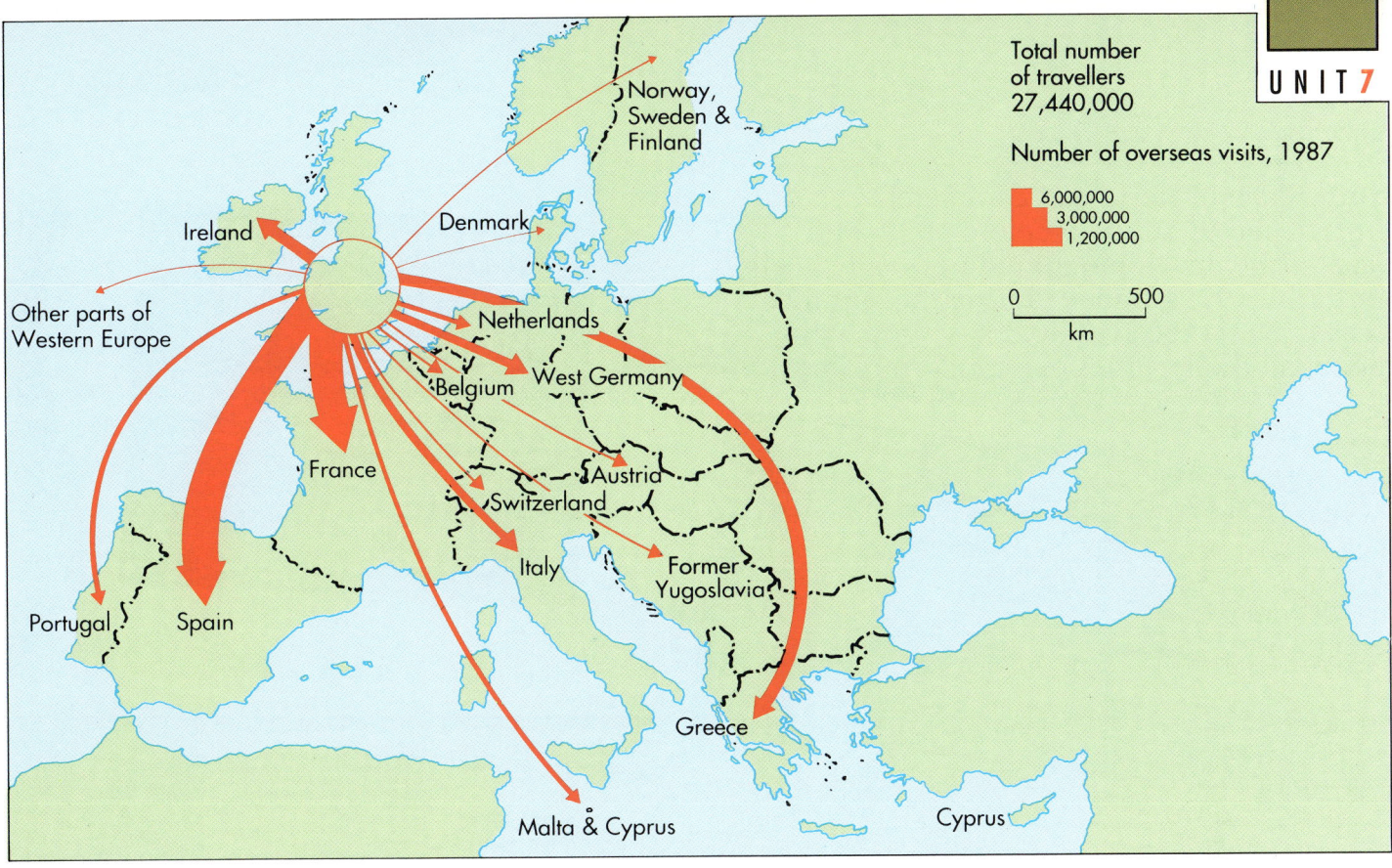

Source: International Passenger Survey, 1987.

7·16 *The destinations of Britons on holiday in Europe in 1987 (before the unification of Germany, and the break-up of Yugoslavia and Czechoslavakia).*

1 Use an atlas to find
a which countries border the Mediterranean Sea and
b which countries have part of the Alps within them.

2 Draw a graph to show the five most popular European countries for British tourists in 1987.

3 Suggest reasons why these countries are popular holiday destinations for British tourists.

The growth of tourism in Spain

Few countries have felt the impact of the tourism explosion as much as Spain, especially in some of its previously more remote and underdeveloped regions.

To get a feeling for Spain *before* the tourist boom, read the following description of the Mediterranean coastal area that we now know as the Costa del Sol. Compare Laurie Lee's description with the image given on page 132 of the Costa del Sol today, in a British holiday brochure.

7·17 *Laurie Lee's experience of Mediterranean Spain in September 1935.*

Then from far out to sea, through the melting mist, would emerge a white-sailed fishing fleet, voiceless, timeless, quiet as air, drifting inshore like bits of paper. But they were often ships of despair; they brought little with them, perhaps a few baskets of poor sardines. The women waited, then turned and went away in silence. The red-eyed fishermen threw themselves down on the sand.

The road to Málaga followed a beautiful but exhausted shore, seemingly forgotten by the world. I remember the names – San Pedro, Estepona, Marbella, and Fuengirola. . . . They were salt-fish villages, thin-ribbed, sea-hating, cursing their place in the sun. At that time one could have bought the whole coast for a shilling.

From 'As I Walked Out One Midsummer Morning' by Laurie Lee, 1969

Nerja: a Costa del Sol resort

The village of Nerja was completely untouched by tourism before 1959, and large-scale development did not really begin until the 1970s. Today, during the summer season Nerja receives 30,000 visitors who enjoy all the facilities of a modern resort.

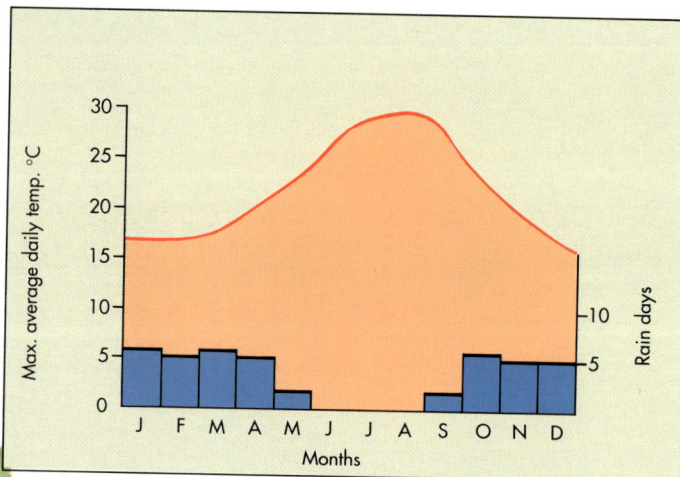

NERJA

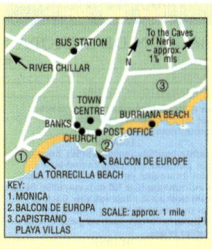

IN OUR OPINION

From the resort's 'Balcony of Europe', a palm fringed promenade built on an outcrop of rock, you'll have a superb view of Nerja's coastline, dotted with inlet coves and beaches.

☑ An elegant Moorish town, undiscovered compared with its famous Costa del Sol neighbour, and still full of character and traditions.

☒ The nightlife, whilst truly Andalucian in flavour, is not as lively or as exuberant as in Torremolinos.

ABOUT THIS RESORT

Beaches — On one side there's the small Calahonda Beach where the local fishermen land fresh fish. A little further on there's the beautiful Burriana Beach — over half a mile of inviting sands, fringed with beach bars and cafés.
Eating Out — A selection of cafés and restaurants, some full of character.
Nightlife — Plenty of bars and four or five discos to choose from.
Shopping — Its intriguing little streets are lined with shops and boutiques.
Transport — There's a regular service from Nerja to Malaga from where you can reach Granada and Seville.
AIRPORT: MALAGA.
TRANSFER: APPROX 1hr 30mins.

HOTEL MONICA
テテテテ
Nerja

RELAXING

ON THE BEACH

"This extremely well-run hotel has a great situation and good range of amenities."

The modern Hotel Monica has earned praise for its comfortable accommodation, good service, food and fine facilities. It stands just across a small promenade from Torrecilla Beach, about 500 metres from Nerja centre.
● pool; terrace with bar; garden
● 2 bars; 2 lounges
● buffet breakfast; waiter service at dinner; à la carte menu; weekly gala dinner; snack bar; daily Happy Hour
● **FREE** tennis (1 hr/room/day subject to availability); table tennis; pool table; keep-fit
● TV room; daily videos; shops
● evening entertainment
● all bedrooms have satellite TV
● children's pool
Prices: based on half board in a room with two, three of four beds, bath, wc and balcony.
For Supplements and Reductions, see Price Panel.
Hotel Bedrooms: 234
Lifts: 4   Official Rating: ★★★★
Tel No: (52) 521100
CHILD REDUCTIONS UP TO 11 YEARS INCLUSIVE

Hotel Monica: half-board for 7 nights (per adult)

Month	£
January	219
February	248
March	282
April	323
May	335
June	355
July	397
August	406
September	380
October	302
November	242
December	310

7·18 *Part of a holiday brochure advertising the resort of Nerja on the Costa del Sol.*

Source: Thomson Tour Operators Ltd.

1a Draw a line graph to show accommodation prices at the Hotel Monica.
b Compare your graph with the climate graph for Nerja. Try to explain any pattern you find.

2 Write your own description of Nerja for a holiday brochure aimed at attracting British tourists.

3 Look at the table below, which shows data about Nerja, then use the figures to draw three bar charts. Write a short description of each graph.

4 Using the information in this section, *either*
a describe what changes have taken place on the Costa del Sol since Laurie Lee's visit in 1935,
or
b imagine you are Laurie Lee making a return visit to Nerja. Think about the way he might react to the changes. Would he see all the changes as changes for the better? Write down what you think his reaction would be.

Year	% of population employed in			Resident population of Nerja	Number of hotel beds
	Agriculture	Fishing	Tourism and related activities		
Pre-1960	90	10	0	c. 8,000	0
1991	12	3	85	14,373	3,061

Leisure can cost the Earth

As the number of tourists increases each year, so the search for new holiday destinations and leisure activities spreads wider too. In the late 1960s and early 1970s, holiday companies opened up a whole new leisure market – the sport of skiing.

There are now 3,000 ski-resorts throughout the world. Nearly half of them are in Western Europe where high snow-covered mountains are quite close to large and relatively affluent industrial populations.

Of the world's mountain ranges, the European Alps have become the busiest winter sports location.

7.19 *Chiesa, an Italian Alpine ski resort.*

Tourism in the Alps

- 20 million tourists spend their holidays in the Alps each year.

- 30 million tourists take day-trips to the Alps.

- A total of 250 million visitor days are spent there each year.

- 40,000 pistes (ski-runs) cover 40,000 km of mountainsides.

- 14,000 ski-lifts can carry 1.25 million people per hour.

Of the many leisure activities associated with winter sports, the largest single event is the Winter Olympic Games. In 1992 they were held around Albertville in the French Alps, and much controversy was caused because of their impact on the environment.

A chairlift cuts through the trees on a mountainside.

Albertville and the surrounding resorts.

Resorts reap the rewards

The 1992 Winter Olympics in the French Alps has unleashed an avalanche of spending in the region.

A new motorway has been built into Albertville 10 years earlier than originally planned, to cope with the Olympic traffic. The rail service to the region has been upgraded too. The local tourist industry is understandably pleased at both developments.

For years the raw sewage of some 200,000 people has spilled into the River Isère. Now the Games have brought the badly needed waste-treatment plants to the Alpine resorts. Dozens of eye-sores have been cleaned up in readiness for the world-wide TV coverage. For example, disused anthracite mines have been landscaped. Of course there has also been a great deal of building and improvements being carried out to make the region ready for the influx of competitors and visitors. The local tourist industry expects to reap the rewards of new hotels, new restaurants and the world-class sports facilities for years after the Olympics have finished.

From a national newspaper

7·20 *The ups and downs of change in the Alps.*

Alps scarred 'forever' by Olympics

Rare habitats ruined as gas masks are issued

THE WINTER OLYMPICS, which opened yesterday, caused massive environmental damage to one of the world's most beautiful areas before even the first ski touched the slopes.

Scientists say that the mountains around Albertville, Savoy, will never recover from the £1 billion preparations for the Games, which have scraped the soil from fragile alpine pastures to make smooth-running pistes. Officially protected forests have been violated and rare marshland damaged. Residents have had to be issued with gas masks because of the danger from toxic chemicals.

The Games – which will be watched by two billion television viewers worldwide – add one more insult to the injury of the Alps, which are rapidly becoming one of the world's foremost ecological disasters. More than half the trees in this backbone of Europe are affected by acid rain, farmers are abandoning their land and landslides and avalanches are increasing under the impact of 50 million tourists a year.

From The Observer, 9 February 1992

Following the 1992 Olympics, the Winter Games were re-scheduled to take place two years after the Summer Games, starting with the 1994 Games in Lillehammer, in Norway.

Imagine that the ski-resort of Livigno in the Italian Alps has made a bid to host the 1998 Winter Olympic Games. Use the information on page 134 to complete the following activities.

1 Make a list of the reasons why Livigno might wish to host the Games.

2 Design a poster that can be used by the International Commission for the Protection of the Alps to show the likely impact of the Winter Olympics on Livigno. Ideas in the poster should be based on the experience of Albertville.

3 Find out more about some of the impacts you have identified. How could the Winter Olympics be better planned in the future?

▷ Is tourism a passport to development?

Economically developing countries have seen the growth of the world tourist industry, and many see tourism as a way to improve their own level of development.

When the Tourists Flew In

The Finance Minister said
  "It will boost the Economy
  The dollars will flow in."

The Minister of Interior said
  "It will provide full
  and varied employment
  for all indigenes."

The Minister of Culture said
  "It will enrich our life . . .
  contact with other cultures
  must surely
  Improve the texture of living."

The man from the Hilton said
  "We will make you a second
  Paradise;
  for you it is the dawn
  of a glorious new beginning!"

When the tourists flew in
  our island people
  metamorphosized into
  a grotesque carnival
  – a two-week sideshow

When the tourists flew in
  our men put aside
  their fishing nets
  to become waiters
  our women became whores

When the tourists flew in
  what culture we had went out
  the window
  we traded our customs
  for sunglasses and pop
  we turned sacred ceremonies
  into ten-cent peep shows

When the tourists flew in
  local food became scarce
  prices went up
  but our wages stayed low

When the tourists flew in
  we could no longer
  go down to our beaches
  the hotel manager said
  "Natives defile the sea-shore"

When the tourists flew in
  the hunger and the squalor
  were preserved
  as a passing pageant
  for clicking cameras
  – a chic eye-sore!

When the tourists flew in
  we were asked
  to be "side-walk ambassadors"
  to stay smiling and polite
  to always guide
  the "lost" visitor . . .
  Hell, if we could only tell them
  where we really want them to
  go!

Cecil Rajendra

7·21 *The poet's view.*

From 'Bones and Feathers', Heinemann, 1978

1 Discuss what the poet says was promised from tourism, and what he says actually happened.

2 Write down one example from the poem of each of the following impacts

which tourism had on the host society:
- environmental impact
- cultural impact
- economic impact
- social impact.

Tourism in Indonesia

Among the developing countries that regard tourism as an important contributor to development is Indonesia. 'Visit Indonesia Year' aimed to promote the country as a world tourist destination using the slogan 'Let's Go Archipelago'. Of the 13,677 Indonesian islands, Bali is the most popular tourist destination.

Holiday brochures market Bali as an island paradise, with idyllic tropical beaches, lush green rice fields and smiling, gentle islanders living in perfect harmony with their gods and the land.

Over the years Bali has received ever increasing numbers of tourists, especially after Ngurah Rai International Airport was extended in 1969. Ninety per cent of visitors are from Indonesia (non-Balinese), Australia, the USA, Japan and Europe.

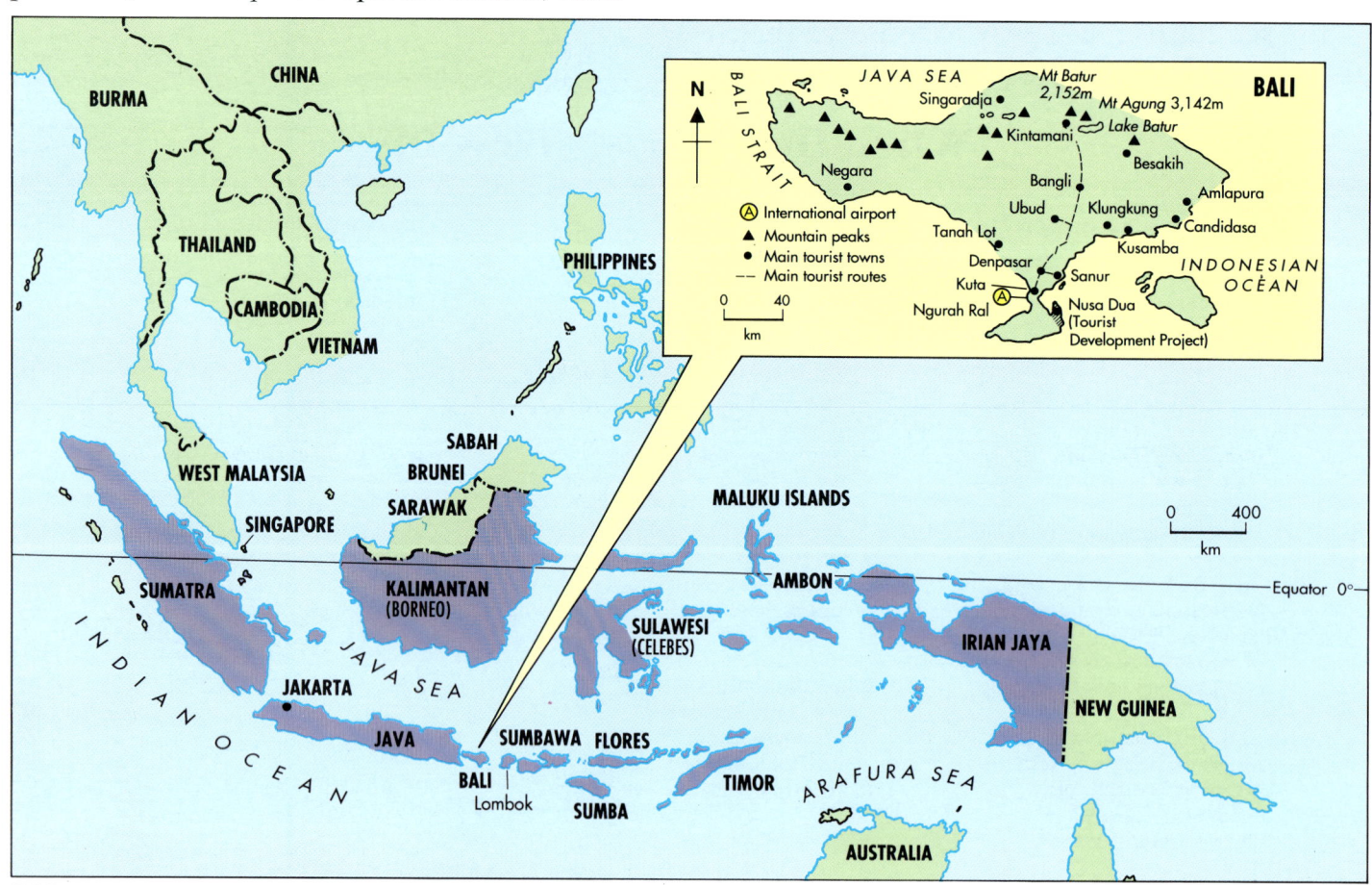

7·22 *According to holiday brochures, the islanders of Bali live in complete tranquillity.*

1a Using the figures in the table below, draw a graph to show the increase in tourist arrivals to Bali, from 1930 to 1994.

b What impacts do the opening of the international airport and promotion of Indonesia appear to have had on tourist arrivals?

Year	Number of tourists
1930	30,000
1969	47,000
1975	276,000
1978	400,000
1982	550,000
1985	650,000
1994 (est.)	2,000,000

c Write out the following sentences using the correct choice of words.
Few/Most visitors to Bali are from other parts of Indonesia and *developed/ developing* countries. *Not many/Many* Australian and Japanese tourists visit Bali because it is *far/close* by jet plane.

2 What do you think the Indonesian government hoped to achieve by promoting the country as a tourist destination?

A working holiday on Bali

In an effort to promote tourism in Bali, the island's Tourist Officer has invited young people from overseas to visit during the school holidays.

Andrew has gone to stay with a Balinese schoolboy, Wayan Tantri, in the coastal village of Candidasa. He has recorded his working holiday in a diary, and included photographs and cuttings.

Candidasa's largest hotel, built to match the traditional Balinese architectural style.

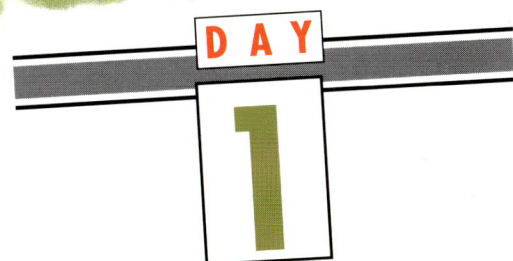

DAY 1

Wayan gave me a tour of Candidasa and the local area. We began by climbing a hill overlooking the village. Looking back it was almost impossible to tell that there was a village below at all. Wayan explained that the Balinese authorities have made it illegal to build above the height of the palm trees.

Looking out to sea we could see 8 large concrete groynes projecting from the beach. Wayan told me that the coral from the reef 400 metres off-shore has been used as building material for hotels in the south of Bali. With much of the reef gone the local beach is being eroded by the waves.

We left the tranquillity of the hills and walked into Candidasa through the coconut groves. The single street where all the tourist amenities are located was surprisingly quiet too. Wayan explained that most of the large hotels have been confined to the south of Bali, near the airport, to avoid development all over the island. It's a shame though that even the quiet street in Candidasa has neon signs and advertising hoardings. Amazing to think that all of this has sprung up since 1983!

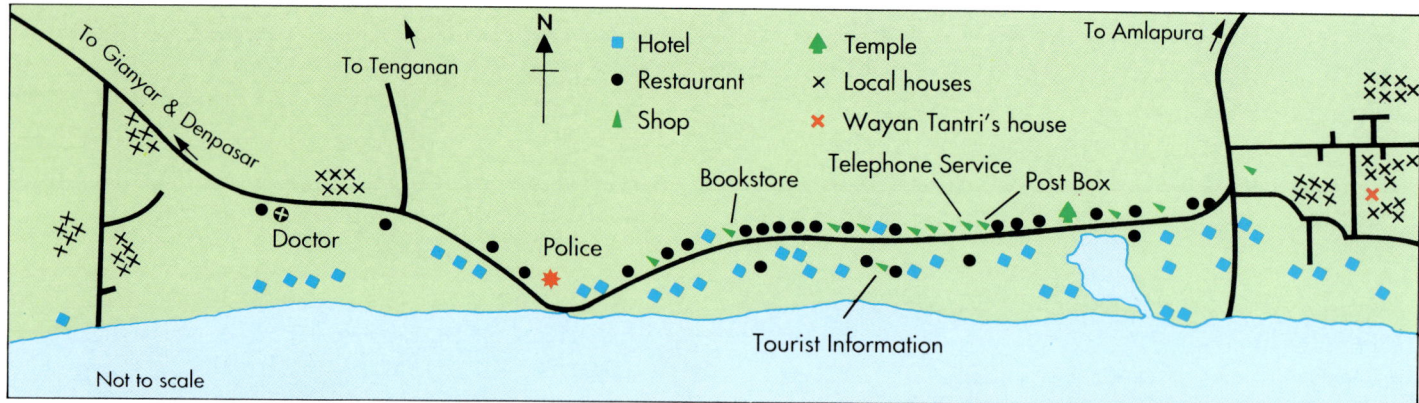

Candidasa, tourist resort.

CANDIDASA

From Denpasar, Klungkung, or Pandangbai, take a *bemo* headed for Amlapura and get off about one km past the turnoff to Tenganan, where you start to see *losmen* signs popping up everywhere. Though this is the fastest-growing beach resort area on Bali, unfortunately the beach itself is not big enough to accommodate all the people they hope to bring in during the next few years. The rhythm and attitude of the locals is noticeably more downbeat than Kuta, though there are signs of creeping Kuta-ism. Consisting basically of *losmen* and restaurants, and the businesses that service them, the locals have moved out to the edge as the resort takes over the central strip. Outsiders from Amlapura and Ubud are starting up all the businesses. Only one business, the Kelapa Mas, is owned by a native of the area. Local color can be found in the wings of Candidasa, however, in the form of a few *warung*. Still, Candidasa is a nice relatively unspoiled getaway if you want white sand, gentle waves, fine dining; few sellers, and nothing to do but soak up the sun by day and lull yourself to sleep with the sound of crickets and crashing surf by night. It's the type of place where you think you'll stay 2 days but end up staying 6!

bemo = bus
warung = foodstall
losmen = guest house

From 'Indonesian Handbook', an American guidebook by Bill Dalton, 1989

1a Make two lists, of the well planned and less well planned environmental impacts of tourism on Candidasa. Give each list an appropriate title.
b For one impact from each list, say *either* why it was well planned *or* how it might have been better planned.

2 Draw a cartoon highlighting some of the impacts you identified in question (1). Give your cartoon a caption.

A new sign at the village temple in Candidasa.

This morning Wayan took me to the main village temple. He explained that anyone was welcome to visit but they were expected to be suitably dressed. Unfortunately some tourists in the past have not shown any respect for the local customs. I bought a beautiful sarong and temple scarf to wear at the temple and to keep as souvenirs.

Before coming to Bali I had heard about the beautiful dancing that was performed at festivals. So I really enjoyed seeing Wayan's sister K'tut among the dancers at the village hall in the evening. Afterwards she told me that although the performance was specially arranged for tourists, it does mean that traditional music and dance are not now dying out.

DAY 3

Up early this morning to join Wayan and his sister K'tut at work. During the school holidays Wayan does part-time work at one of the 'losmen'. He showed me a newspaper advertisement and said that he was saving up to go to university in a few years' time so that he could get a similar job.

Vacancies

A Bali resort hotel is seeking the following:
1 Assistant Housekeeping Manager
2 Assistant Front Desk Manager
3 Assistant Reservation Manager
4 Assistant Sales Manager
5 Front Office Cashier

All candidates must have the following qualifications:
1 Minimum 3 years' experience in a similar position.
2 Good proficiency of written and spoken English.
3 Hardworking and adaptable with initiative.
4 University graduates are preferred.

Please send your application accompanied with CV and 2 recent photographs to:

Personnel Manager
P.O. Box 2056
Kuta – Bali

K'tut works long hours as a maid in a hotel doing the cleaning and washing for the tourists. I don't think she gets paid much for her hard work. I was shocked at the way some tourists didn't even say 'thank you' to K'tut in their own language (let alone in Balinese) after she'd cleaned their rooms.

As I visited the tourist accommodation I noticed that a lot of the furniture and fittings came from outside Bali and Indonesia. Even the manager was an Australian!

1a Under the heading 'Economic changes', make a list of the differences between life in Candidasa *before* and *after* the arrival of tourism.
b Do you think that both Wayan and K'tut would consider the changes in your list as changes for the better? Explain your answer.
2 What examples can you find to show that profits from tourism are not staying in Candidasa?

DAY 4

Got up this morning at 3 am to go fishing with Made, Wayan's brother. Before the tourists came, fishing was his only income, but as there wasn't much to spend any money on it didn't matter. A few years ago Made also made money by collecting coral from the reef. That is now illegal, so he takes tourists out with him fishing or out to the reef to snorkel. He gets about 10,000 rupiah (3,300 rupiah = £1) from tourists for each trip. With the money he has bought an outboard motor for his 'prau' (fishing boat), and a moped, and is hoping to buy a TV when the house is connected to the electricity supply. We caught five snapper which Made sold to a restaurant for 100 rupiah each.

Decision-making activity

Now that your working holiday is drawing to a close, you need to report back to the Tourist Officer. She wants you to help her produce a Code of Conduct for Better Tourism. This should include what you consider are the most important rules to be followed to achieve the most positive results from future tourist developments. You need to consider the views of
• the host society • the visitors
• the planners • tourist businesses.

In small groups, discuss your ideas for a Code of Conduct. Then make a final Code of Conduct for Better Tourism, with no more than 12 rules, ready for presentation to the Tourist Officer.

The leisure society

In the developed world we enjoy many activities for pleasure. We live in a society with the free time, money and mobility to do so. City dwellers in Britain, for example, demand access to the countryside at weekends. National parks have been created to balance this demand for leisure with the need for conservation.

Tourist industry

A huge tertiary industry has grown up to service the need of our leisure society to travel for pleasure. Package-deal holiday companies have made it possible for millions of people to travel abroad every year for their holidays. The tourist industry has now reached every part of the world.

Tourism and development

Tourism is the world's fastest-growing industry. To many developing countries tourism appears to be a good way to improve local standards of living.

Indonesia has been keen to sell its sunny climate, sandy beaches and exotic cultures. In Candidasa some people have benefited more than others from tourism.

Impacts of tourism

These are the effects of our holiday-making. Impacts are usually felt most at holiday destinations. They are often measured in terms of change to the host *society* (which includes the culture and the economy) and the *environment*. The way of life for the people of Nerja in Spain, for example, has been changed with the arrival of tourism.

Some impacts of tourism are positive: for example, the creation of many jobs in hotels and shops in Nerja. Other impacts are negative: there is now congestion in the streets of Nerja and pollution on the beaches of the Costa del Sol, for example.

Future planning

The future of tourism depends on the way in which it is planned and managed. Without planning, those things that tourists go on holiday to experience may disappear.

Brazil – a Land of Opportunity?

a Tropical rainforest.

b The drier region of the south-east.

8·1 *Landscapes of Brazil.*

a The inner city.

Look at the four pictures on page 141 and on this page. Then work in small groups.

1 Divide a large sheet of paper into four sections.

2 Fill each section with words that describe each of the four photographs.

3 Each person in the group should now be responsible for one photograph. Write a description of that photograph using the words that the group have suggested.

4 Write down what you think has caused these areas to become like they are. Is it the climate (temperature, rainfall)? Is it the people who live there (size of population, what they do for a living)?

b A poor suburban area.
8·2 *São Paulo, Brazil's industrial centre.*

8·3 *Brazil.*

Setting the scene

All the pictures you have just described are part of one country – Brazil. It is a large country of enormous wealth. It was for many centuries attractive to European merchants and adventurers. Today the official language of Brazil is Portuguese – Christopher Columbus, the first explorer ever to cross the Atlantic, was sponsored by the Portuguese.

Brazil is a country that is developing fast and is aiming to be a world leader in the 21st century. In 1973, President Hedia stated that Brazil wanted to be 'a free, prosperous, strong and independent nation with a prominent place among other great nations' (*Brazil Herald*). In 1992 Brazil became a prominent place: the leaders of over 140 countries

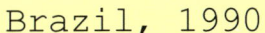

Brazil, 1990	
Area	8,457,000 km²
Population	153.3 million
Population density	16 per km²
Birth rate	27 per 1,000
Death rate	8 per 1,000
Natural increase	1.9%
Doubling time	36 years
Infant mortality	63 per 1,000
Life expectancy	65 years
% population under 15	35%
% population over 65	5%
Urban population	75%
Gross national product per person	$2,550 [1989]

What on Earth can Rio achieve?

8·4 *US Senator Gore and Indian chiefs at the Earth Summit, Rio de Janeiro, 1992.*

gathered in Rio de Janeiro for the first-ever 'Earth Summit'. This was a conference at which proposals for helping countries to beat poverty *and* protect the environment were discussed.

Brazil can be divided into five geographical regions: northern, centre-west, north-east, south-east and southern. Each area has different characteristics. It is a land of sharp contrasts. It is very rich and very poor, very wet and very dry, very advanced and very traditional.

Brazil has many resources that could help its development. There are natural resources like gold, diamonds, offshore oil and gas, bauxite, minerals and huge reserves of iron ore. There are also products like sugar, coffee, rubber, beans, orange juice and more recently beef and soya products, which are exported all over the world. For many years Brazil has also been a world leader in the export of woods from its huge rainforest areas.

8·5 *Clearing the rainforest.*

8·6 *Cattle ranching in Brazil.*

The arguments about the destruction of the rainforests have often centred on Brazil because of its vast forested areas that are gradually being removed as a result of shifting cultivation, cattle ranching, and the export of hardwoods like mahogany.

In the past, raw materials were exported to be refined or manufactured in other countries. Now the government encourages industries to refine products before they leave the country, so that more of the profits go to the Brazilians. Building things like oil refineries and iron and steel plants has been a priority for the Brazilian government. New industries are growing and factories are using ideas developed abroad to produce goods cheaply and efficiently.

Spectacular large-scale schemes like the Itaipú Dam scheme have been a feature of Brazil's history. These schemes have been successful, but often not for the poorer people. The Trans-Amazonian Highway was a similar project, designed to cut right across the rainforest. Some parts are in daily use, whilst other areas have not stood up against the effects of the tropical climate and have been washed away by heavy rains.

But the business people in Brazil are still keen to help industry with new, bigger and better schemes.

8·7 *Itaipú Dam on the Rio Parana.*

Just over three years since the idea was first mooted, one of the most ambitious railway construction schemes of the century – 1,178 km in its first phase – is now all set to come off the drawing board.

From 'Oil and Fats International 2', 1991

144

How can Brazil develop?

1 Work in small groups. Each group's task is to produce a large information sheet or poster to show what the different areas in Brazil are like. You are not to use any words except for the main headings. Discuss what methods of presentation you will use. Each person should have a specific job to do.

2 Appoint one person to be the Editor, who must design a general introduction to the country and be in charge of layout and headings. Everyone else must be responsible for presenting information on *one* of the regions. You may use the pictures on pages 141–144 and the information boxes that follow to help you.

3 When the individual pieces of work are ready they must be put together to form a complete 'picture' of Brazil. Display your information sheets/posters around the classroom.

8·8 *Aspects of Brazil.*

Parana

3% of the land area of Brazil

Population: 9 million

Problems: Lack of a skilled workforce

Advantages: Agricultural resources, tax incentives to investors

Industries: Cars, textiles, timber, furniture, farm produce

Developments: Advanced technology industries, manufacturing

a Parana.

São Paulo

3% of the land area of Brazil

Population: 33 million

Problems: Traffic congestion, pollution and crime in urban areas

Advantages: Established multinational companies and commercial centre

Industries: Food processing, vehicles and electrical goods

Developments: Telecommunications, pharmaceuticals, port

b São Paulo.

Espirito Santo

0.6% of the land area of Brazil

Population: 2.5 million

Problems: Small area, mountainous interior

Advantages: Agricultural and mineral resources, good transport links

Industries: Steel, oil, minerals, wood, coffee

Developments: Steel production, wood pulp

c Espirito Santo.

North-east

18% of the land area of Brazil

Population: 40 million

Problems: Drought, lack of food, people moving away

Resources: Agricultural and mineral resources, large workforce

Industries: Food, textiles and chemicals

Developments: Deepwater port, steel plant, oil refinery, tourism

d North-east region.

So how is Brazil to develop in the future? Can Brazil use its huge advantages of natural resources and varying geographical regions and turn them into riches for *all* the people? In this unit we will examine this issue by looking at three key questions:

▷ What is it like living in a large city in Brazil?

▷ How can developing transport help Brazil in the 21st century?

▷ How shall we develop trade links with Brazil?

▷ **What is it like living in a large city in Brazil?**

A class of children in an international school in the Brazilian city of São Paulo were asked this question. Many of these children move around often because their parents work in multinational companies or with foreign governments. So many of the children had not been in Brazil very long when they were asked this question, and they were able to give their first impressions of the city. Here are some of their answers.

I came to Brazil last November from Japan and I've found a lot of differences between Brazil and Japan. The buildings are older and Brazil is more dangerous than Japan.

People in Brazil are active, noisy and most of them are friendly.

São Paulo is quite a busy place to live in – it never stops, and you're always doing something no matter what.

Most people go to the cinema and shopping centres at the weekends. At night people go to discos and some might go to clubs.

Life in São Paulo is good. There are a lot of things to do, like going to the malls or the parks, movies, discos and many shops. It's good living here, but there are some very serious dangers like robbery in the streets.

Jane was a teacher in London until recently, when she decided she needed a change of scene and a chance to see a bit more of the world. She is now at the international school in São Paulo, and she describes here how her life has changed since she moved.

I have a maid who comes in two half-days a week and does all the cleaning, washing, washing-up and ironing. Everyone I know here has a maid – all apartments have areas for maids to work in, including separate toilets. It is one way of providing jobs for people who may otherwise be out of work.

Depending on who you talk to in a Brazilian city like São Paulo, you might get a very different picture of life there. Whilst some people would talk about having a maid and having a good time, others would talk about how hard life is and what a struggle it is to keep enough food on the table and to make enough money to live on.

As in all economically developing countries, there are big differences in the lifestyle of Brazilian people, depending on where they live and the kind of job they can do. Many people had the choice of remaining in the countryside in extreme poverty with the threat of death from starvation or disease, or moving to the already overcrowded cities to live in slum dwellings or *favelas*. The council say they will throw the slum-dwellers off the land, but they have nowhere else to go.

1 Imagine that someone from another country wanted to know about your way of life. Write down what you would tell them.

2 What differences do you notice between what you have written and what the children in Brazil wrote (page 147)?

3a Write down the names of two places that you know nearby – perhaps in the same town or not too far apart – that are different from the place where you live.
b Explain how they are different. Try to write down some reasons why you think they are different.

4 'Life in Brazil depends on where you live.' Would you say the same about life in Britain?

8·9 *The extreme stress of living in poverty in Brazil's 'favelas' results in the break-up of family life. Thousands of orphans – 'street children' – make a rough and dangerous living in the cities. They have no other choice.*

Life in Brazil depends on where you live. It is a hard life for some people and a very good, calm life for others. We're the unlucky ones. Life on the streets is dangerous. The police don't like us. We don't go to school. All of us have friends who have just disappeared.

8.10 *Middle-class houses in São Paulo.*

The wealthy people in São Paulo tend to live in the centre of the city. Many people choose to live in high-rise luxury apartments, although some people still prefer to live in houses. These richer people have servants and enjoy many luxuries.

Many of the pupils at the international school come from wealthy families. Here two pupils describe the apartments in which they live.

Many people still move into the cities from the countryside. They often leave behind poverty, malnutrition or even starvation. The droughts often last much longer than the people can cope with; they have no food and can grow no crops without the rain.

People leave the countryside with no money. When they arrive in the city, the only places where they can afford to live are the favelas, which lie on the edge of many of the big cities in Brazil. Here people make houses for themselves from any suitable materials they can find – cardboard, sacking, sheets of metal. From here they often travel into the city to look for a job in the factories in the industrial areas. There is a chance that children in the favelas can go to school, but if their families stay in the countryside they are likely to remain unemployed and the children may never have the chance to go to school.

Some things in São Paulo are the same, whoever you are and wherever you live. There are approximately 12 million people living in this city, and sometimes it seems as if they all want to be on the move at the same time. There is often incredible congestion on the roads of São Paulo, and then it's quicker to walk! A journey by car can be twice the distance of the direct route because of all the one-way roads.

I live in a luxury apartment at the moment but we're going to rent a house in Jardin, Paulista. That's a suburb of São Paulo, close to the school, and we will move into it soon. The rent is going to be very high – US$2,700 per month.

Our house has a tiny front garden and back garden. In the front are two tall, thin palm trees, and occasionally small green parrots come to eat the fruit. Our house is double-storeyed and has three bedrooms and four shower rooms.

8·11 *Traffic flowing into the centre of São Paulo.*

8·12 *Pollution in São Paulo.*

There are many new industries in São Paulo but there are not yet as many laws to prevent pollution as there are in countries like Britain. When Jane, who had been teaching in London, first came to live in São Paulo, she was quite ill.

> I picked up a fairly vicious dose of 'flu – in this polluted city they breed strong bugs, and there's no fresh air to blow them away!

The climate varies from region to region but in São Paulo the weather can vary from day to day as well as from season to season.

> Here in the city it is very hot. Today, for example, the heat is suffocating! But when it rains it really rains. We have floods and sometimes power cuts.

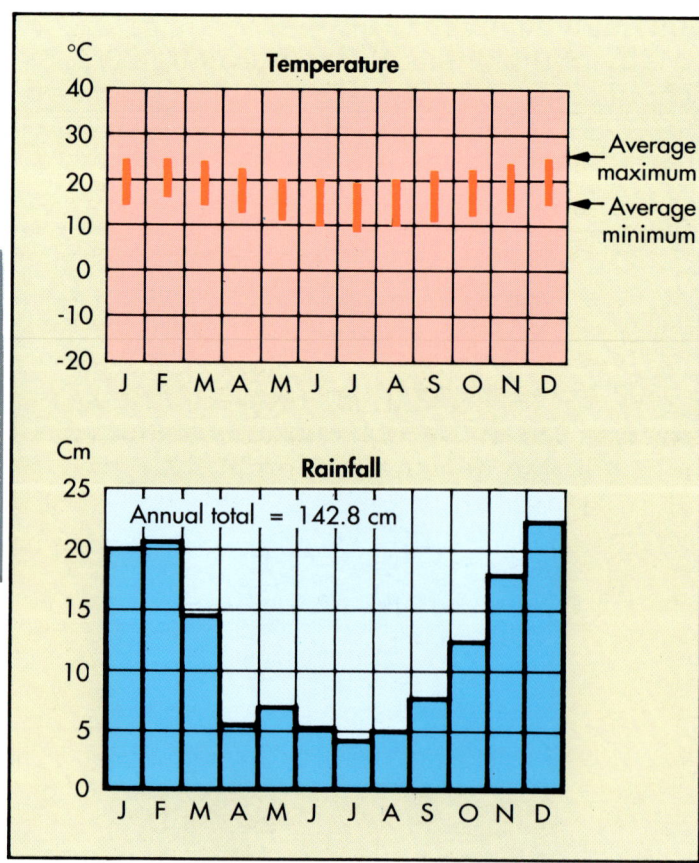

8·13 *Climate graphs for São Paulo.*

▶ How can developing transport help Brazil in the 21st century?

8·14 *Transport network in Brazil.*

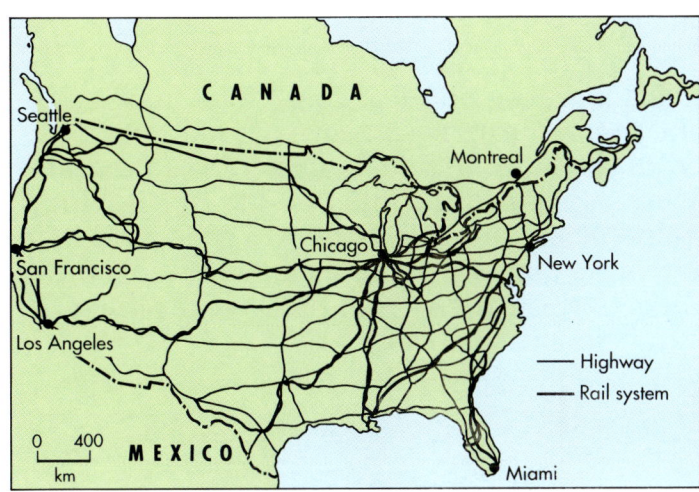

8·15 *Transport network in the USA.*

60 – 120 cm high

8·16 *The soyabean plant.*

New transport links are being planned in Brazil. They are huge, expensive and important new projects that could help Brazil's industries grow. Goods must be transported quickly and cheaply to the ports if they are to be exported and earn money from abroad. Better transport links should help Brazil to develop economically.

Better transport is needed to help export the farm products from the large farms, especially soya beans. In the USA, for example, only 5% of the factory price of soya is made up of transport costs. In Brazil transport costs make up about 50% of the price. Better transport links would cut transport costs for Brazilian farmers, and so make them more competitive. Transport in the soya-growing region of Brazil is poor, which is why the price of soya is so high compared with the price in the USA, which has a network of interstate highways.

Soyabean supplies feed for animals, food for humans, and raw materials for industry. It is one of the world's most useful and cheapest sources of protein. It contains about 40% protein – beef and fish contain much less, so in many countries people choose to eat soya protein instead of the more expensive forms of protein.

Soyabeans grow well in the mid-west of Brazil because physical conditions there suit the plant. Farmers can earn a good return from the land. It is such a useful crop that if it were developed it could provide Brazil with a valuable export. The beans are processed in crushing plants, which need good transport links to the coast so that the soya products can be exported easily.

8·17 *Soyabean-crushing plants in Brazil.*

Some people think that better transport links would provide more opportunities for employment. People would benefit as goods are moved around more quickly. Many think they would be good for the country as a whole.

But this is a controversial issue. Some people may lose their land, others complain that they will not be near enough to the transport for it to help them. Others still hold the opinion that the money needed for such schemes would be better spent in other ways that would benefit the people more directly.

There are three ways in which farm produce can be transported in Brazil: by road, by rail or by river. There are of course advantages and disadvantages with each of these forms of transport.

Roads

Until now roads have been the most popular form of transport used by soya farmers. Road transport is expensive but it is fast and can be arranged to suit the individual farmer or can be organised to accommodate several farmers. Roads to the major farm areas are usually passable all year round, but the roads to some of the more remote areas can sometimes become impassable. Pollution is a problem along the roads, and so is congestion at the port.

Roads

- 2,000 km of roads planned to be extended or strengthened.
- Investment needed: $250 million to improve the network significantly.
- Estimated 50,000 people employed during construction.
- Estimated 10,000 people employed to drive lorries and subsequently for maintenance.
- Estimated 2,000 people made homeless.
- Increase in air pollution estimated at 10%.
- Loss of rainforest estimated at 10,000 km².

Rivers

- 1,038 km of waterways made accessible.
- Investment needed: $97 million.
- Estimated 40,000 people employed during construction.
- Estimated 7,000 people employed to run boats and subsequently to do maintenance.
- Estimated 500 people made homeless.
- Increase in water pollution estimated at 2%.
- Loss of forest estimated at 1,000 km².
- Additional bonus of electricity generated at the dams.

Railways

Railways do already exist in Brazil, and there are plans to develop more in the future. A train can carry large amounts of produce at one time, creating less pollution than road transport. It is also very fast. Once the soya is loaded it goes directly to the port. The problem with railways is that they do not go directly to every farm, so the soya must first be loaded onto lorries and is then unloaded onto the trains.

Railways

- 1,718 km of new rail track planned.
- Investment needed: $620 million for Stage 1.
- Estimated 60,000 people employed during construction.
- Estimated 88,000 people employed to run trains and subsequently for maintenance.
- Estimated 1,000 people made homeless.
- Increase in air pollution estimated at 3%.
- Loss of forest estimated at 5,000 km².

Water

Water transport along the navigable rivers of Brazil is a fairly new way of transporting farm produce to the ports, as it is only recently that some rivers have become passable. It is by far the cheapest way of transporting goods but it is also the slowest and again the soya must be transported by road from the farm to the river.

The map below shows those areas of the country where the different forms of transport would be developed.

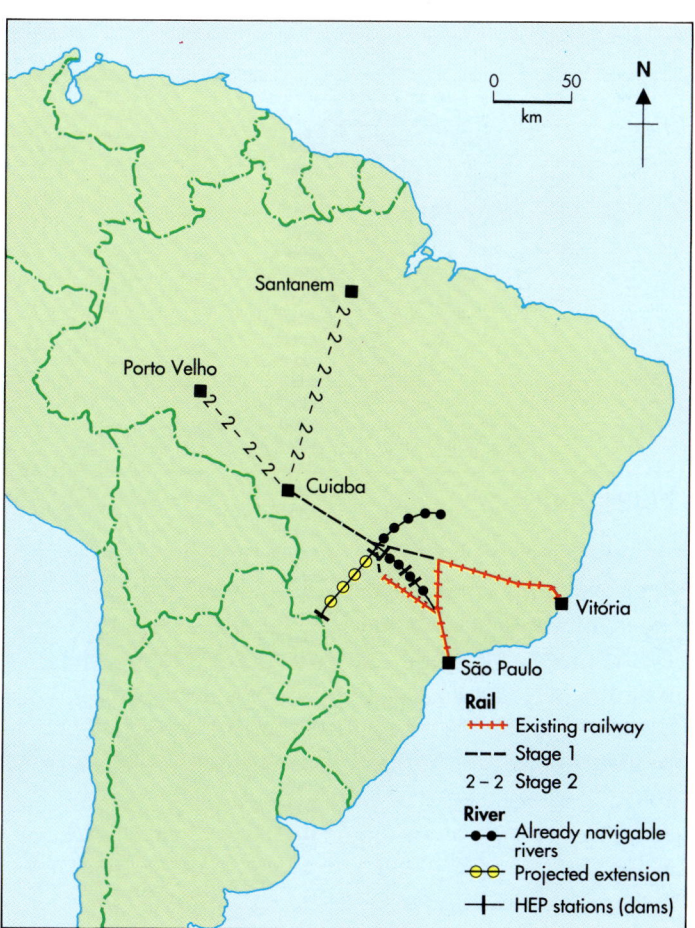

8·18 *Rail and river transport developments in Brazil.*

Here are some of the views of the people involved.

Lorry driver

The roads are the best way to develop transport in Brazil. We've established routes all over the country already and all we need to do is add some extra lanes to the existing roads, and to extend some of them, to have a really quick, efficient and comprehensive network across the country. People are used to using road transport, and I don't think they'll want to change. Lots of jobs will be created, and that's what people are bothered about round here. They need work so that they can feed their children and keep body and soul together.

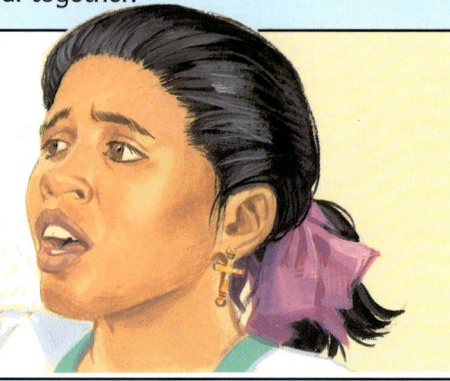

Maria Sanchez

You simply can't allow any more traffic on this road – it's bad enough as it is. Lorries thunder past my house day and night, and it would only get worse. I have young children and they might be in danger from the road being so close to our house. The baby would never get any sleep. The pollution will get worse. All the dust and fumes would be breathed in by my children, and they say that the lead and fumes can harm the children over a long period of time. They even say they might make the road wider, in which case I'd lose my home altogether!

Environmentalist

By far the best of these schemes is to develop the rivers. They are the natural way to transport goods, and the minimum amount of harm will be done to the environment. The best thing about it is that they can make electricity at the same time. This could mean that homes that have never had electricity will be able to afford it for the first time, instead of only the very rich people being able to have such basic things as electric lighting and heating. The least amount of pollution will be caused by river development, and the loss of our precious forest areas will be kept to a minimum if this scheme is chosen.

Politician

Schemes like this are just what the government have been wanting. They'll help all our industries in the inland areas of our country, and we don't have to find all the money for it. Private investors are taking the responsibility for a lot of the finances. It will lead to the increases in Brazil's exports that the government has been encouraging over the past years. This means that the economy of the whole country will develop and the standard of living will improve for everyone. More jobs will be created and we can become the main influence in South America as far as international relations are concerned.

Work in small groups. Imagine that you are neighbouring farmers in the mid-west of Brazil. Consider all the points of view that have been put forward, and decide which form of transport you and the other local farmers will use.

1 *Each person* must select a different mode of transport and write down a list of the advantages and disadvantages of that form of transport.

2 *As a group*, decide which form of transport you will use. Explain your reasons for choosing that one.

▷ How shall we develop trade links with Brazil?

Often, the links which developed countries had with Brazil were very beneficial for the developed country, but not for Brazil. Often Brazil sold its raw materials at low prices to the developed country and in return bought manufactured goods at very high prices. This meant that Brazil acquired a very large debt, and many Brazilian people had a very low standard of living, and never saw the manufactured goods that were imported – only the very wealthy people could afford them.

As a country Brazil now faces a dilemma. It has to create a balance between two courses of action.

Firstly, it needs to develop its own industries so that it can provide jobs for many people and improve their standard of living. It also means that goods from these industries can be bought by other Brazilian industries more cheaply than they would be if they were imported. More jobs will be created, and more income made for the country as a whole.

Secondly, links need to be established with foreign or *transnational companies* (TNCs) so that Brazilian goods can be sold to them, and other goods can be produced. Brazil would also learn more about technical products and methods from foreign companies. However, it is not easy to attract foreign industry to a country like Brazil. If the conditions are not just right, a transnational company will look for another country where conditions are more favourable.

Mike Croghan is the Sales and Marketing Manager of the transnational company Unilever in São Paulo. He says there are problems for his company operating in Brazil. He wrote down these problems in a list.

- Rapid inflation and economic uncertainties
- Poor infrastructure for telecommunications and transport
- High taxation
- Inefficient Brazilian industries
- Import restrictions
- Unnecessary bureaucracy
- Social problems and poverty
- Corruption
- No agricultural policy to ensure a steady supply of materials, for example Soya beans.

155

All these problems together mean that it is very difficult to operate a business in Brazil, because of all the laws, restrictions and administration that people must go through to make sure they are not operating illegally.

However, there are still huge advantages for conducting business in Brazil. One of the most important is *the sheer size and potential of the market*, as the country has such a large population. There are *abundant natural resources*, and also a hard-working *labour force* in São Paulo and the southern states, which has the added advantage of being very cheap. There is also a fast and efficient *banking* system in Brazil.

So how does Mike see Brazil developing in the future?

Although Brazil obviously has a great future, the speed of development will depend mainly on the people in government, and the reforms they can bring about. A strong government with power to carry out long-term economic and social reforms is fundamental to progress. This means improving education, health, law and order, and transport connections: these are the basis for a successful society.

If Brazil could address its problems and truly open up its economy, as some of the other South American countries have done, then it could be a real world force.

Developing industry successfully has many 'knock-on' effects. These can be drawn in a diagram like the one on the right.

1 Mike Croghan is not a Brazilian. He works for a transnational company. When he talks about trade links between Britain and Brazil, he has a certain viewpoint. Try to describe his viewpoint. To do this, work through the following activities:
a Choose the statements below which best describe Mike's priorities (look back to the 'Values and viewpoints' card on page 152). Write out the list in what you think would be Mike's order of priority.
● I want the company (my employer) to be successful.
● I want to provide jobs for as many Brazilians as possible.
● I want to help make Brazil a rich and fair society.
● I would like the company to be able to sell products to Brazil's large market.
● I would like the company to help develop Brazil's large natural resources.
b Find evidence in what Mike says to support your decision about the order of Mike's priorities.

2 Write down how you think the order of priorities would be different for
a a Brazilian government leader
b an unemployed factory worker from a favela in Rio de Janeiro.

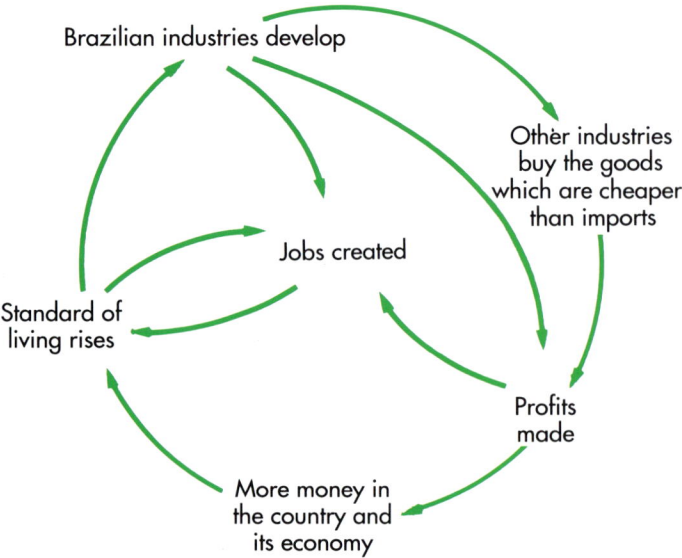

8·19 *A cycle of economic development.*

3 Mike Croghan describes some advantages of developing industries in Brazil. For an alternative viewpoint, study the cartoon and newspaper article below. Explain how some of the problems in the sack may be problems for Brazil. (For example, where will pollution come from? What effects will it have?)

Brazil's leading exports of primary products:

1	Coffee	20% of total
2	Soya	16%
3	Minerals	16%
4	Orange juice	5%
5	Meat	5%
6	Cocoa	5%

The Development Dilemma for Brazil

Back in the 1970s, Brazil was told by the experts of that time that the answer to its problems of poverty was 'economic growth'. The way out, they said, was 'exports'. *Beef* was needed for the huge, growing US hamburger market, and *timber* was needed for the booming construction industries of the world. By a clever piece of thinking, the two were linked – Brazil could produce both!

So in came the *loans* and the machinery, and down came the forests to make way for cattle ranches. Out, too, went the peasant farmers, to make way for the ranches.

All was well: the trees were shipped out, followed by the meat. Together they made a few people very rich. But then came erosion. Brazil's fragile soil – where the forest had been – started to wash away. More loans came to buy fertiliser, and for the factories needed to process the beef and the timber.

Exports grew. For a while the experts – and the banks – loved Brazil and its 'miracle of economic growth'.

More recently, *different* experts have studied the hidden costs of the great export drive. The erosion of the soil and the loss of forest have cost Brazil billions of dollars a year. Hundreds of thousands of ordinary people have lost their livelihood. Poverty has actually increased!

What is Brazil left with? Destroyed land, even poorer peasants, and a few very rich people. The banks also began to turn away from Brazil: the prices for beef and timber fell dramatically during the 1980s, and Brazil found it difficult to repay its enormous debts.

In the 1990s some Brazilians began to realise that the damage was even more serious. Cutting the forest down for beef and timber destroys the *bio-diversity* resource of Brazil (and the world), which is vital for future bio-technology industries.

From a national newspaper

8·20 *Disadvantages of industrial development in a developing country.*

For this activity work in small groups, or pairs. Each group should choose *one* of the activities.

1 *Dilemma* – for Brazilian leaders

You are Brazilian representatives to the next Earth Summit. (You were organisers for the 1992 Earth Summit.) Your concerns are

- to help economic development
- to reduce poverty
- to protect the environment, both locally and globally.

You have been asked to make a speech entitled: 'How the rich countries must change their ways'. Prepare your speech.

2 *Dilemma* – for transnational companies (TNCs)

You are the Chairperson of a TNC with its world headquarters in New York. Each year your company spends several billion dollars on investment – in factories, machines, transport, and so on. Your concerns are

- to make money *responsibly*, and stay in business
- to give all people and countries a fair chance
- to protect the global environment.

Brazil, an indebted nation with vast natural resources, has offered you very good terms to help its development: there will be no planning controls on building and waste disposal, and no local taxes!

You can foresee the future shareholders' meeting at which you receive applause for your skill at raising the profits of the company yet again. What do you do?

The 1992 Earth Summit

Full title: United Nations Conference on Environment and Development

Biggest conference ever held – over 120 world leaders including some of the richest (e.g. USA) and poorest (e.g. Sudan) nations of the world.

Venue: Rio de Janeiro, Brazil

Aim: To change the way the world's economies are run in the name of sustainable development. Sustainable development is the ability of one generation to hand over to the next at least the same amount of resources.

'We have to look at Earth in a businesslike way. Earth, the business, is going bankrupt. Much of the income we produce is not really income at all; it's just running down our savings.'

Maurice Strong, UN organiser

Proposals for saving the planet

These were the main proposals made at the 1992 Earth Summit. Not all were signed or agreed by all countries. The USA had to be persuaded to go to the conference and once there found little it could agree with.

1 *World agreement on climate:* New laws to reduce energy use in order to avoid future global warming.

2 *World agreement on 'bio-diversity':* New laws to protect habitats.

3 *Declaration on forests:* To end large-scale forest destruction for timber (logging).

4 *Agenda 21:* An action plan to help countries develop and at the same time care for the environment.

Proposals like these cause problems both for rich countries and for poor countries. But it is the richer countries of the world that may have to change most.

Brazil, an economically developing country

Brazil has a high dependency on exporting *primary products*. There are huge differences in wealth between the few very rich people and the mass of poor people; these *wealth inequalities* are growing wider. Much of Brazil's environment is under threat of *degradation*, and Brazil owns a large share in the $429,174,000,000 *debt* (1991) that South America owes to the wealthy, advanced industrialised nations.

Urban growth

A common problem in economically developing countries is very rapid growth of the towns, especially the larger ones, caused by the migration of poor or dispossessed people from the countryside.

Low-paid jobs are available in manufacturing industry. This is an attraction. But wages are not enough to pay for a comfortable urban lifestyle: rents are too high, food is expensive, bus fares prohibitive. So hundreds of thousands of people risk living in dangerous and illegal conditions in the favelas.

Transport development

Brazil is a large country – the fourth largest on Earth. All large countries face similar problems: holding the country together and enabling the whole nation to develop. Brazilians see the building of a *transport infrastructure* – roads, railways and waterways – as vital for economic development.

Like all aspects of space and society, though, the idea is controversial. What is more important: extremely expensive transport schemes, or the protection of 'unspoilt' environments and human cultures?

Transnational companies (TNCs)

We all depend to a large extent on the activities of TNCs. They bring in energy (e.g. BP), transport (e.g. Ford), entertainment (e.g. Sony), and a range of foods and household products (e.g. Unilever). They also provide jobs and bring economic development to many places across the world.

On the other hand, economically developing countries do not appear to benefit from TNCs in a simple, straightforward way. The companies pay low wages, they *exploit* resources for profit rather than for the development of the country, and they are difficult to control.

International trade

Many economically developing countries rely on primary products for export. They depend on importing machinery and other manufactured goods. This puts them in a 'trade trap', selling cheap and buying dear.

Earth Summit

Brazil hosted the 1992 Earth Summit. Was this the beginning of a long process by which the countries of the world began to solve the twin problems of wealth inequality and environmental destruction?

The events of your lifetime will form part of the answer to this question.